Makin' Numbers

Makin' Numbers
Howard Aiken and the Computer

edited by I. Bernard Cohen and Gregory W. Welch
with the cooperation of Robert V. D. Campbell

The MIT Press
Cambridge, Massachusetts
London, England

© 1999 Massachusetts Institute of Technology

Set in New Baskerville by Wellington Graphics.

Printed and bound in the United States of America.

Library of Congress Cataloging-in-Publication Data

Makin' Numbers : Howard Aiken and the Computer / edited by I. Bernard Cohen and Gregory W. Welch with cooperation of Robert V. D. Campbell.
    p.   cm.— (History of computing)
   Includes bibliographical references and index.
   ISBN 0–262-03263–5 (hc : alk. paper)
   1. Aiken, Howard H. (Howard Hathaway), 1900–1973. 2. Computer engineers—United States—Biography. 3. Computers—History.
   I. Cohen, I. Bernard, 1914–. II. Welch, Gregory W.
QA76.2.A35M35   1999
004'.092—dc21
[b]   98-43964
                                         CIP

Contents

Appendixes

Preface

The title *Makin' Numbers* comes from a favorite phrase of Howard Aiken's, one he used to express his happiness when his "computing machines" were busy finding numerical solutions to problems or producing tables of various sorts.

Part I, the heart of the book, contains technical discussions of Aiken's first machine, the IBM Automatic Sequence Controlled Calculator (also known as the Harvard Mark I), and of the improved version, Mark II.

Aiken played an important role in the development of higher education in the area we now know as computer science. In part II, two of his Ph.D. students, Frederick Brooks Jr. and Peter Calingaert, offer reminiscences and evaluations of that program and provide information concerning the atmosphere of Aiken's Computation Laboratory at Harvard. Their chapters are supplemented by Gregory Welch's critical discussion of the Harvard environment and the problems of establishing the new discipline there.

The reminiscences by Grace Hopper, Richard Bloch, Anthony Oettinger, Maurice Wilkes, and Henry Tropp in part III reveal many aspects of Aiken's complex personality.

For part IV, Gregory Welch has produced a selection of extracts from Aiken's talks, based on transcripts available in the Harvard University Archives.[1]

The chapters by Bloch, Brooks, Campbell (on Mark I), Hopper, and Wilkes are based on presentations at the Pioneer Day celebrations of Aiken and his machines. Organized by Robert Campbell, Richard Bloch, and I. Bernard Cohen, these celebrations were held during the National Computer Conference of the American Federation of Information Processing Societies at Anaheim, California, in May 1983. Tapes and transcripts of all the papers given are available in the Harvard University Archives, and transcripts are on deposit in the Charles Babbage Institute at the University of Minnesota.

The chapters by Bashe, Calingaert, Tropp, and Welch were written for this volume.

Readers who are interested in Aiken's life and career beyond the scope of this volume may consult *Howard Aiken: Portrait of a Computer Pioneer* (MIT Press, 1999).

1. A few of these have been published in proceedings of meetings or conferences; however, in his whole career he wrote only one paper expressly for publication: "Trilinear Coordinates" (*Journal of Applied Physics* 8 (1937): 470–472). An article attributed to him, titled "The Future of Automatic Computing Machinery," was published in *Nachrichten Technische Fachberichte* (4 (1956): 31–35), but an editorial note explains that this was "a transcription from magnetic tape" of a lecture and that it had "not been revised by the author." Aiken was also listed as a co-author, with Grace Hopper, of three articles on "The Automatic Sequence Controlled Calculator" (*Electrical Engineering* 65 (1946): 384–391, 449–454, 522–528); I gather that Grace Hopper prepared these, which were essentially an abridged version of the Manual of Operation for Mark I. Aiken's other publications are forewords or afterwards to symposiums. His doctoral dissertation, Theory of Space Charge Conduction (officially dated "Harvard University, January 1, 1939"), is available in the Harvard University Archives; there is no record of his having published any portion of it.

Acknowledgments

Grateful acknowledgment is made to the Charles Babbage Institute for a small grant which was very useful in getting this project started, and to the Richard Lounsbery Foundation of New York City for its generous support which made the completion of this volume possible. Gregory Welch, as well as writing his own important contributions, helped in editing the whole volume. Robert Campbell's good counsel and deep understanding of the founding period of the computer age have enriched almost every stage in the preparation of the volume.

—I.B.C.

Introduction

Howard Aiken's primary contribution to the computer is often said to have been his first machine, the IBM Automatic Sequence Controlled Calculator or Harvard Mark I. Conceived in 1937 while Aiken was completing his doctoral dissertation at Harvard, that machine was brought into existence by IBM engineers (primarily Clair Lake, Francis Hamilton, and Benjamin Durfee) working in close consultation with Aiken, who made many trips from Harvard to IBM's plant in Endicott, New York, where the machine was constructed.[1] Aiken's original proposal for this machine (1937), a landmark document in the history of the computer, is presented here for the first time exactly as Aiken wrote it. It is followed by Robert Campbell's description of Mark I, its mode of operation, and some of the travails that were encountered in getting it to perform without error. Campbell is uniquely qualified to write this presentation: he was the person designated by Aiken to oversee the completion of the machine when Aiken was called to active duty in the Navy.

Aiken was fortunate in obtaining the services of Campbell, a Harvard graduate student in physics. Few graduate students would have had the ability and the maturity to see this project to conclusion at IBM's plant and have then been able to take charge of an untried machine of enormous complexity and work out the many problems that arose when it was finally put into operation. The magnitude of Campbell's contribution is not generally recognized. He was in charge of the actual operation of the machine, and he had to learn "on the

1. When the machine was formally presented to Harvard, in August 1944, a conflict arose between Aiken and IBM (notably Thomas J. Watson Sr.) over the allocation of credit for the invention. For details see *Howard Aiken: Portrait of a Computer Pioneer* (MIT Press, 1999)—henceforth cited as *Portrait*.

job" how to run it. In fact, he had to invent a system of programming. In this endeavor he had no manual or guide; there was no body of experience to guide him in charting wholly new pathways in large-scale computing. Furthermore, Campbell had to convert mathematical problems into programs for the machine. Here again there was no precedent, no source of experience on which he could call. A true pioneer, Campbell was on his own, an adventurer into a new world.

The author of the companion chapter, Charles Bashe, was for many years a member of IBM's engineering and development staff. Bashe established IBM's corporate technical history project and was the senior author of *IBM's Early Computers* (MIT Press, 1985). He has drawn on the company's archives to produce an account of the development of the new machine from an "inside IBM" viewpoint.

The author of the chapter on the programming of Mark I, Richard Bloch, is also uniquely qualified. During 1944 and 1945, as an officer in the Navy and a member of Aiken's staff, he served as Mark I's principal programmer. Both Howard Aiken and Grace Hopper recalled that Bloch was the only person they had ever known who could write a computer program in ink; Bloch holds this to be an exaggeration.

Campbell later became the chief engineer in charge of the construction of Mark II. Mark II, commissioned by the Navy in 1945, was designed as an "improved" Mark I, without radically new componentry. Indeed, the design of Mark II may be taken as an index of the needs for improvement that became apparent after Mark I was put into regular operation. Of course, by 1946, when Mark II was becoming operational, ENIAC (the Electronic Numerical Integrator and Computer, built for the Ballistic Research Laboratory at Aberdeen, Maryland, by the University of Pennsylvania's Moore School of Electrical Enginering) had been completed and had demonstrated the enormous advantages of electronic elements over relays in large-scale computing machines. The path to the future thus was shifted from Aiken's machines to ENIAC. Although Mark I and Mark II continued to do useful work for many years (which may be taken as an index of the increasing national need for computing services), their technology was obsolete.

In his chapter on Mark II, Campbell makes a point about the significance of that machine that is usually not appreciated: Aiken had shown that he could produce a complex calculating machine on his own, with his own staff.

Aiken's two later machines, Mark III and Mark IV, did not affect the mainstream of computer development and accordingly are not given separate chapters. Mark III was innovative in that it had a magnetic drum memory, Mark IV in that it had magnetic cores. Mark III had about 4500 vacuum tubes, some solid-state devices, and some relays; Mark IV was all electronic. Mark III had a built-in coding machine that greatly simplified the work of programmers; Mark IV was notable for having a device that would automatically translate equations into coded instructions.[2]

Aiken often expressed hostility toward the use of pure binary systems in computer design. However, Mark II used binary-coded decimals, and Mark III was based on an apparently original version of the binary-coded system. The subject of number systems arose in the autumn of 1973, when Henry Tropp and I interviewed Aiken as part of an oral-history program sponsored by the American Federation of Information Processing Societies and the Smithsonian Institution. Tropp, who had been trained in the history of mathematics at the University of Toronto, was on the mathematics faculty of Humboldt State University. During the interview, he discussed with Aiken some features of number systems, especially the use of a ternary system. Tropp's presentation of this topic discloses a hitherto unknown facet of Aiken's thinking about computer design.[3]

One of Aiken's most significant contributions to the computer was the founding of a program for graduate instruction in what is known

2. For further information concerning Mark III and its mode of operation, see "Description of a Magnetic Drum Calculator," in *Annals of the Computation Laboratory of Harvard University* 24 (1949). On Mark IV, see the privately printed Manual of Operation for the Harvard Magnetic Drum Calculator (Mark IV), prepared by Norman B. Solomon and dated July 1957. (Robert Campbell plans to add his copy to the Aiken files in the Harvard University Archives.) Various features of Mark IV are described in detail in the 60 volumes of Progress Reports to the Air Force (AF-1–AF-60), dating from May 1948 to December 1960, and in the 123 supplementary Problem Reports of the Computation Laboratory (April 1949–July 1959). These are available in the Gordon McKay Library of the Division of Engineering and Applied Science of Harvard University, in Pierce Hall.

3. Tapes and a transcript of the interview conducted by Tropp and Cohen in September 1973 are on deposit in the Harvard University Archives and in the archives of the Charles Babbage Institute. Unfortunately, the audio tape recording is imperfect, and there are some gaps in Aiken's explanations. On the number system used in Mark III, see appendix E of *Portrait*.

today as computer science. He recognized early that there was a need for individuals with a knowledge of applied mathematics, experience in programming, and an understanding of computer design and function. Toward this end, he established at Harvard, in 1947–48, a graduate program leading to a master's degree and to a doctorate.[4] The first two doctorates were officially awarded in Engineering Sciences and Applied Physics; the later doctorates were in Applied Mathematics. Aiken's doctoral students and the titles of their dissertations are listed in an appendix.

Aiken's teaching and training program is the subject of part II. Frederick Brooks Jr., who received his Ph.D. at Harvard under Aiken, describes what it was like to work with Aiken and to belong to the Harvard "Comp Lab." Brooks, later celebrated for his role in the design and production of the IBM System/360, eventually became chairman of the Department of Computer Science at the University of North Carolina. In his recollections of the Harvard Comp Lab, he stresses the importance of the interaction among graduate students and the close relationship they had with Aiken. Peter Calingaert completed his Ph.D. under Aiken, then stayed on for a while as a member of the staff of the Comp Lab. He later joined Fred Brooks at the University of North Carolina. Calingaert's presentation recreates the atmosphere of the classroom and conveys a student's impression of Aiken as a teacher. (When interviewing Aiken, I had asked him whether Tropp and I might see his lecture notes; Aiken replied that he had always destroyed his lecture notes at the end of each year, so that he would not be tempted to repeat his lectures. Accordingly, Calingaert's recollections are of special importance.)

Gregory Welch and Adam Rabb Cohen have studied the career of Howard Aiken and the early history of computer science at Harvard. In his contribution to this volume, Welch considers some of the difficulties Aiken faced in trying to establish computer science at Harvard. These difficulties were due in part to the administration's unwillingness to provide adequate financial support, but Welch believes that there was also a general hostility toward a seemingly "applied" activity at Harvard, an essentially humanistic university. In a concluding note to Welch's chapter, Cohen addresses Harvard's relative

4. Instruction in large-scale machine computing had been inaugurated a year earlier at Columbia under the direction of Wallace Eckert and Herbert Grosch.

decline as a center of the new computer science in relation to the rise of other centers, notably MIT and Stanford.

The five recollections in part III present Aiken at three different points in his career. Grace Hopper, later famed for her roles in the invention of the compiler and in the creation of COBOL, describes her encounter with Aiken and the Comp Lab when she was a newly commissioned Navy officer.

Richard Bloch's reminiscences of Aiken during World War II give us a glimpse of Aiken's human side, particularly his concern for the health and well-being of his staff. All too often this aspect of Aiken's personality is overlooked.

Anthony Oettinger looks at Aiken from the viewpoint of a student, colleague, and successor who was privy to many of "the Boss's" insights and prejudices.

Maurice Wilkes's chapter is important because it sets Aiken's reputation in a European context. Wilkes reveals Aiken's delight in intellectual jousting and his respect for worthy opponents; he also gives us a look at Aiken at a turning point in his career, when he no longer was at the forefront of the new science of the computer.

Henry Tropp concludes part III with a portrait of Aiken in his seventies, just weeks before his death. Tropp introduces us to a side of Aiken that was markedly different from the "difficult" personality that so many people remember. The solicitous concern that Tropp describes here reminds us of Aiken's care for his staff, invoked by Richard Bloch in an earlier chapter.

—I.B.C.

The Name "Mark I"

In this volume, Aiken's first machine is generally referred to as "Mark I" or "the Harvard Mark I." ("Harvard" is sometimes added to distinguish this machine from the "Mark I" at the University of Manchester and from other "Mark I" machines.)

When the machine was installed at Harvard, in 1944, Thomas J. Watson Sr. ordered a stainless steel sheath with "IBM AUTOMATIC SEQUENCE CONTROLLED CALCULATOR" emblazoned on it. Before long, however, Aiken introduced the name "Mark I," and ever since then the machine has been generally known by this simpler designation (although some IBMers still prefer "IBM ASCC").

When Aiken began to use the name "Mark I" is not known with certainty, but he probably began to do so when he was asked to design and construct a second machine; in accord with Navy practice and general technological usage, the new machine was "Mark II."

"Mark I" is the designation most commonly used today. In recent years, high-ranking IBMers, including historian Emerson Pugh and former CEO Thomas J. Watson Jr., have adopted it.

Makin' Numbers

Howard Aiken, circa 1960.

Introducing Howard Aiken
I. Bernard Cohen

Howard Aiken, one of the founders of the computer age, earned his place in the historical record with several sets of achievements. In addition to designing and building four giant calculators (i.e., computers) and inaugurating an academic program in what is now known as computer science, he was one of the first to explore applying the new machines to business purposes. Aiken also advocated the use of computers in language translation, in textual and historical analysis, and in economics. Much in demand as a speaker in the United States and in Europe, he constantly urged introduction of the computer into new areas of research and action. His contributions to the computer age also include the symposia he organized to bring together those who were designing and constructing computers or planning new applications.

Aiken's primary claim to a notable place in history was Mark I, his first machine, built from his specifications by IBM engineers. This machine showed scientists and engineers that a machine could solve complicated mathematical problems by being programmed to execute a series of controlled operations in a predetermined sequence—and that it could do so without error.

A strong man both physically and in quality of mind, Aiken forced Harvard, an essentially humanistic university, to take a leading role in the new science of computing. Some of his visions of the future so outran the course of events that his predictions were often not validated for decades. He was strongly opposed to the new concept of the stored program, especially the mixing of program or instructions with data in the same store, and he had a basic dislike of pure binary number systems.

Aiken was born in Hoboken, New Jersey, in 1901, and was educated in the public school system of Indianapolis and at the University of

Wisconsin, where he earned a bachelor's degree in electrical engineering. After ten years of success as an electrical engineer, he entered graduate school at Harvard University, where he received a Ph.D. in communication engineering in 1937.

While engaged in his thesis research, Aiken encountered mathematical problems that were difficult and time-consuming. The standard way of solving such problems was to use an electro-mechanical desk calculator and published tables of mathematical functions. Aiken decided that such laborious calculation could be mechanized, and he set about designing a machine for that purpose. Such a machine, he envisioned, could solve pressing problems in science and in engineering, and also in the social sciences.

Aiken brought his proposal to the Monroe Calculating Machine Company. Monroe's chief engineer, George Chase, was enthusiastic, but the company was not interested in the venture. Aiken then turned to IBM, where his project immediately won the support of the chief engineer, James Wares Bryce, and soon the backing of the chief executive officer, Thomas J. Watson Sr. IBM agreed to convert Aiken's proposal into working reality. (Aiken's original proposal is included in this volume, and the inside story of the construction of the new machine is set forth in Charles Bashe's chapter.) Between 1939 and 1941, Aiken regularly traveled to Endicott, New York, to give specifications to the IBM engineers who were building the new machine and to help in designing its circuits.

In 1941, when Aiken (an officer in the Naval Reserve) was called to active duty, he designated Robert Campbell to be his deputy in the final stages of completion of the machine. Campbell was also responsible for making test runs and for programming and running the first problems in the spring of 1944, after the machine was moved to Harvard University and became operational. Aiken returned to Harvard to take command later that spring, and for the duration of the war the Harvard Computation Laboratory was run as a Navy project.

Although Aiken's relations with the IBM engineers and executives began cordially, the harmony was broken by events at the "dedication" of Mark I at Harvard in August 1944.[1] A major cause of the rupture

1. The ceremony at which Thomas J. Watson of IBM officially presented the ASCC/Mark I to Harvard has become generally known as "the dedication." This expression does not occur in the Harvard news release, nor does it appear in the official invitations. During the planning of the event, however,

was a news release, issued by the university, that presented Aiken as the primary, if not the sole, inventor of the new machine. (For details concerning this episode, and other aspects of the construction and operation of the new machine, see *Portrait;* also see the chapter by Frederick Brooks in the present volume.)

Many important problems were assigned to the new machine during World War II, one concerning the evaluation of steel for use in constructing ships, one concerning magnetic mines, one concerning ray tracing for a new type of telephoto lens for the Air Force, and one concerning implosion. (The implosion problem was brought to the Comp Lab by John von Neumann. The programming that was done for it is described in Richard Bloch's chapter.)

Toward the end of the war, the Navy commissioned Aiken to design and construct a second machine much like Mark I but incorporating some improvements based on the experience with Mark I. The second machine, Mark II, was sent to the Naval Proving Ground in Dahlgren, Virginia, where it had a productive and useful life. (Details concerning the architecture of Mark II will be found in Robert Campbell's second chapter.)

Aiken and his staff went on to build two more machines. Mark III went to Dahlgren to join Mark II. Mark IV, built for the Air Force, remained at Harvard. Whereas Mark I and Mark II were relay machines, Mark III used some vacuum tubes and some solid-state devices. Mark IV was all electronic, using selenium solid-state devices that were later replaced by ones made of germanium. Both Mark III and Mark IV introduced some important novel features, notably the use of magnetic drum storage and (in Mark IV) the use of magnetic core memories. One of the truly innovative elements of Mark III was an automatic coding device intended to simplify the work of the programmer and to avoid human error in the writing of programs. Although Aiken's designs included some conditional branching, none of these machines used stored programs. Aiken insisted, narrowly and strictly, on maintaining the separate identities of data and instructions.

---

the word "dedication" occurs in a letter from T. J. Watson to J. B. Conant (21 April 1944) saying that he was "looking forward to being present at the dedication." There is some reason to suppose that this designation originated with Watson.

Mark III used a binary-coded number system that appears to have been devised by Aiken especially for that machine. In this system, there were binary digits with "weights" or values 2*, 4, 2, 1. Aiken used the symbol 2* to distinguish the value 2 in the fourth place from the 2 in the second place. As may be seen in table 1, translating the decimal digits 0 through 9 into the hybrid binary system is straightforward. Aiken found that this number system had certain advantages when used in Mark III. Examination of the 1 component alone indicated whether a digit was odd or even. Similarly, examination of the 2* component alone indicated whether a digit was less than 5. The nines complement of each decimal digit was obtained by inverting the binary digits, 0 and 1. Three of the four binary components had the same weights as in the binary number system, which permitted many of the simple properties of that system to be retained.

Technologically, Mark I can be considered an important break-through because it embodied a convincing demonstration of the possibility of large-scale, error-free, complex calculations in a programmed sequence. Public announcements concerning the new machine made it known to scientists and engineers the world over that the computer age was dawning. ENIAC, an all-electronic machine at the University of Pennsylvania's Moore School, had not yet been completed, and wartime secrecy and ignorance veiled the British Colossus machines and those of Konrad Zuse in Germany. But Aiken's later machines did not have the "state of the art" character of Mark I, even though Mark III and Mark IV had very innovative features. Mark II performed useful work, but its relay technology was made obsolete by the advent of ENIAC. By and large, Aiken's four machines did not greatly influence the developing computer technology.

In retrospect, Aiken's most important and lasting contribution to the computer may have been his boldness in adapting the computer to data processing at a time when computers tended to be giant, one-of-a-kind number crunchers designed primarily for solving complex mathematical problems. Another area in which his innovations were of real significance was in computer education. Dismayed by the very small number of mathematical doctorates being produced, he was aware of the need for applied mathematicians interested in numerical analysis and in the use of the new computers. Accordingly, he established at Harvard, beginning in 1947, a graduate program in what we know today as computer science, leading to M.A. and Ph.D. degrees. The first degrees were in Engineering Science and Applied Mathemat-

Table 1.1
Decimal digits: 2*, 4, 2, 1 notation.

	2*	4	2	1
0	0	0	0	0
1	0	0	0	1
2	0	0	1	0
3	0	0	1	1
4	0	1	0	0
5	1	0	1	1
6	1	1	0	0
7	1	1	0	1
8	1	1	1	0
9	1	1	1	1

ics, but later ones were in Applied Mathematics. Not all the computer scientists produced by Harvard in those early days went through Aiken's program, however. An Wang, a computer scientist and the founder of the company that bore his name, completed his doctorate at Harvard under Ronald King in communication engineering.

Although Aiken appears to have produced at Harvard the first Ph.Ds in computer science, it is a matter of record that the Harvard program was not the first to offer instruction in that area. Columbia University announced a program of "Instruction and Research in the Watson Computing Laboratory" for "the Winter Session 1946–1947"— that is, a year before Aiken's program got underway at Harvard. Two courses were offered in that inaugural year at Columbia, one of which, on "Numerical Methods," was taught by Herbert R. J. Grosch, a specialist in astronomical and general machine calculation. The aim of that course was to "acquaint research students in science with the theory and practice of computation." It was stressed that "special reference [would] be made to methods useful with recently developed calculating equipment." The other course at Columbia, "Machine Methods of Scientific Calculation," was taught by Wallace J. Eckert with the assistance of Robert Rex Seeber Jr. (who had spent a year with Aiken and Mark I) and Lillian Hausman (the Watson Laboratory's "Tabulating Supervisor").

In 1961, Aiken took advantage of Harvard's policy of allowing faculty members to retire at age 60 with full benefits. He moved to Fort

Lauderdale and was given an appointment at the University of Miami that did not require any teaching. He then became an entrepreneur, taking over ailing businesses and nursing them back to financial health, whereupon they were sold. He also kept up his computer activities, serving as a consultant to Lockheed and to Monsanto. His final contribution in the computer field was a means of encrypting data. He died in 1973 in St. Louis while on a consulting trip.

Aiken received numerous honors and awards. His eminence in the computer field was recognized in 1964 when he received the American Federation of Information Processing Societies' inaugural Harry Goode Award for "outstanding achievement in the field of information processing." He was awarded honorary degrees, medals, and decorations both at home and abroad. In retrospect, some old computer hands consider his greatest contribution to have been the training of young individuals who then went on to advance the art and science of the computer, to design and construct new computers at the cutting edge of the technology, and to direct the new departments of computer science at various universities.

I

Aiken's Machines

Proposed Automatic Calculating Machine
Howard Aiken

Aiken's formal proposal exists in at least three copies, all of which are in the Harvard University Archives. One is among the university's presidential papers for 1938, in a file marked "Physics." The last page is dated "Jan. 17, 1938" and signed "Howard Aiken." This document was formally transmitted to President J. B. Conant on 7 February 1938 by Professor Harry Mimno, whose covering letter to Conant mentions that Aiken had already made contact with IBM and that the two IBM engineers who had studied the proposal had found "the fundamental design" to be "practical." Clearly, the proposal had been written and submitted to IBM before 7 February 1938. On the first page of this copy of the proposal, someone has noted: "This was written in 1938 before construction was started." A second copy, in the files of the School of Engineering, is signed "Howard H. Aiken" in ink and is dated, in Aiken's hand, "January 17, 1938"—the same date that is on the president's copy, and apparently the day on which Aiken officially presented this document to the dean. The third copy, in the Aiken files, is dated in pencil, in a hand that has not been identified, "November 1937." The text is identical in all three copies, which were reproduced by some process. (They are not carbon copies of a typewritten document.)

There is evidence that the date of composition of Aiken's proposal (as opposed to the date of formal transmission to the president and the dean) is 1937. Reference 8 at the end of chapter 2 of the Manual of Operation for Mark I reads "H. H. Aiken, Proposed Automatic Calculating Machine (1937), p. 18, (privately distributed)." Since there are several references to IBM technology, it would appear that this proposal was prepared for Aiken's first contact with IBM: his meeting with James Wares Bryce, IBM's chief engineer, which took place in early November 1937. It does not seem likely that Aiken would have referred to IBM and its machines if this proposal had been written for his unfruitful meeting with George Chase of the Monroe Calculating Machines Company on 22 April 1937. (For additional evidence to support the date 1937, probably November, see Portrait.)

This landmark text, published in IEEE Spectrum *in 1964, has been reprinted from that journal in the three editions of Brian Randell's source book on the antecedents and early history of the computer,* The Origins of Digital Computers—Selected Papers *(third edition: Springer-Verlag, 1975). It has also been reprinted in at least one other anthology of writings concerning the computer.*

The original text, reprinted here, differs in some features from the version printed in IEEE Spectrum *and reprinted in Randell's anthology, which contains a number of alterations. For example, some of Aiken's mentions of "International Business Machines" were changed to "IBM machines." "International Business Machines Company," however, was kept as Aiken wrote it, as was Aiken's "MacLauren" for the name of the Scottish mathematician Colin Maclaurin. Also altered were the paragraphing and numbering of some sections. One section in which many alterations were made is titled "Present Conception of the Apparatus." In the edited version, the numbering and paragraphing were altered, and the displayed and numbered lists lost their numbers and were converted into paragraphs. The text printed below reproduces Aiken's original document word for word.*

I. Historical Instoduction

The desire to economize time and mental effort in arithmetical computations, and to eliminate human liability to error, is probably as old as the science of arithmetic itself. This desire has led to the design and construction of a variety of aids to calculation, beginning with groups of small objects, such as pebbles, first used loosely, later as counters on ruled boards, and later still as beads mounted on wires fixed in a frame, as in the abacus. This instrument was probably invented by the Semitic races and later adopted in India, whence it spread westward throughout Europe and eastward to China and Japan.

After the development of the abacus no further advances were made until John Napier devised his numbering rods, or Napier's Bones, in 1617. Various forms of the Bones appeared, some approaching the beginning of mechanical computation, but it was not until 1642 that Blaise Pascal gave us the first mechanical claculating machine in the sense that the term is used today. The application of his machine was restricted to addition and subtraction, but in 1666 Samuel Moreland adapted it to multiplication by repeated additions.

The next advance was made by Leibnitz who conceived a multiplying machine in 1671 and finished its construction in 1694. In the process of designing this machine Leibnitz invented two important devices which still occur as components of modern calculating machines today; the stepped reckoner, and the pin wheel.

Meanwhile, following the invention of logarithms by Napier, the slide rule was being developed by Oughtred, John Brown, Coggeshall, Everard, and others. Owing to its low cost and ease of construction, the slide rule received wide recognition from scientific men as early as 1700. Further development has continued up to the present time, with ever increasing application to the solution of scientific problems requiring an accuracy of not more than three or four significant figures, and when the total bulk of the computation is not too great.

Particularly in engineering design has the slide rule proved to be an invaluable instrument.

Though the slide rule was widely accepted, at no time, however, did it act as a deterrent to the development of the more precise methods of mechanical computation. Thus we find the names of some of the greatest mathematicians and physicists of all time associated with the development of calculating machinery. Naturally enough, these men considered mechanical calculation largely from their own point of view, in an effort to devise means of scientific advancement. A notable exception was Pascal who invented his calculating machine for the purpose of assisting his father in computations with sums of money. Despite this widespread scientific interest, the development of modern calculating machinery proceeded slowly until the growth of commercial enterprises and the increasing complexity of accounting made mechanical computation an economic necessity. Thus the ideas of the physicists and mathematicians, who foresaw the possibilities and gave the fundamentals, have been turned to excellent purposes, but differing greatly from those for which they were originally intended.

Few calculating machines have been designed strictly for application to scientific investigations, the notable exceptions being those of Charles Babbage and others who followed him. In 1812 Babbage conceived the idea of a calculating machine of a higher type than those previously constructed, to be used for calculating and printing tables of mathematical functions. This machine worked by the method of differences, and was known as difference engine. Babbage's first model was made in 1822, and in 1823 the construction of the machine was begun with the aid of a grant from the British Government. The construction was continued until 1833 when state aid was withdrawn after an expenditure of nearly £ 20,000. At the present time the machine is in the collection of the Science Museum, South Kensington.

In 1934 George Scheutz of Stockholm read the description of Babbage's difference engine and started the construction of a similar machine with the aid of a governmental grant. This machine was completed and utilized for printing mathematical tables. Then followed several other difference engines constructed and designed by Martin Wiberg in Sweden, G. B. Grant in the United States, Leon Bolleé in France, and Percy Ludgate in Ireland. The last two, however, were never constructed.

After abandoning the difference engine, Babbage devoted his energy to the design and construction of an analytical engine of far higher powers than the difference engine. This machine, intended to evaluate any algebraic formulae by the method of differences, was never completed, being too ambitious for the time. It pointed the way, however, to the modern punched card type of calculating machine since it was intended to use for its control perforated cards similar to those used in the Jacquard loom.

Since the time of Babbage the development of calculating machinery has continued at an increasing rate. Key driven calculators designed for single arithmetical operations such as addition, subtraction, multiplication, and division, have been brought to a high degree of perfection. In large commercial enterprizes, however, the volume of accounting work is so great that these machines are no longer adequate in scope.

Hollerith, therefore, returned to the punched card formerly used by Babbage and with it laid the ground work for the development of tabulating, counting, sorting, and arithmetical machinery such as is now widely utilized in industry. The development of electrical apparatus and technique found application in these machines as manufactured by the International Business Machines Company, until today many of the things Babbage wished to accomplish are being done daily in the accounting offices of industrial enterprizes all over the world.

As previously stated, these machines are all designed with a view to special applications to accounting. In every case they are concerned with the four fundamental operations of arithmetic, and not with operations of algebraic character. Their existence, however, makes possible the construction of an automatic calculating machine specially designed for the purposes of the mathematical sciences.

II. The Need for More Powerful Calculating Methods in the Mathematical and Physical Sciences

It has already been indicated that the need for mechanical assistance in computation has been felt from the beginning of science, but at the present time this need is greater than ever before. The intensive development of the mathematical and physical sciences in recent years has included the definition of many new and useful functions nearly all of which are defined by infinite series or other infinite processes. Most of these are inadequately tabulated and their application to scientific problems is thereby retarded.

The increased accuracy of physical measurement has made necessary more accurate computation in physical theory, and experience has shown that small differences between computed theoretical and experimental results may lead to the discovery of a new physical effect, sometimes of the greatest scientific and industrial importance.

Many of the most recent scientific developments, including such devices as the thermionic vacuum tube, are based on nonlinear effects. Only too often the differential equations designed to represent these physical effects correspond to no previously studied forms, and thus defy all methods available for their integration. The only methods of solution available in such cases are expansions in infinite series and numerical integration by iterative methods. Both these methods involve enormous amounts of computational labor.

The present development of theoretical physics through Wave Mechanics is based entirely on mathematical concepts and clearly indicates that the future of the physical sciences rests in mathematical reasoning directed by experiment. At the present time there exist problems beyond our ability to solve, not because of theoretical difficulties, but because of insufficient means of mechanical computation.

In some fields of investigation in the physical sciences, as for instance in the study of the ionosphere, the mathematical expressions required to represent the phenomena are too long and complicated to write in several lines across a printed page, yet the numerical investigation of such expressions is an absolute necessity to our study of the physics of the upper atmosphere, and on this type of research rests the future of radio communication and television.

The roots of transcendental equations and algebraic equations above the second degree can be obtained only by successive approximations, and if an accuracy of ten significant figures is required the numerical labor in many cases may be all but prohibitive.

These are but a few examples of the computational difficulties with which the physical and mathematical sciences are faced, and to these may be added many others taken from astronomy, the theory of relativity, and even the rapidly growing science of mathematical economy. All these computational difficulties can be removed by the design of suitable automatic calculating machinery.

III. Points of Difference between Punched Card Accounting Machinery and Calculating Machinery as Required in the Sciences

The features to be incorporated in calculating machinery specially designed for rapid work on scientific problems, and not to be found in calculating machines as manufactured for accounting purposes, are the following.

1. Ordinary accounting machines are concerned entirely with arithmetical problems, while machines designed for mathematical purposes must be able to handle both positive and negative quantities.

2. For mathematical purposes, calculating machinery should be able to supply and utilize a wide variety of transcendental functions, as the trigonometric functions; elliptic, Bessel, and probability functions; and many others. Fortunately not all these functions occur in a single computation; therefore a means of changing from one function to another may be designed and the proper flexibility provided.

3. Most of the computations of mathematics, as the calculation of a function by series, the evaluation of a formula, the solution of a differential equation by numerical integration, etc., consist of repetitive processes. Once a process is established it may continue indefinitely until the range of the independent variables is covered, and usually the range of the independent variables may be covered by successive equal steps. For this reason calculating machinery designed for application to the mathematical sciences should be fully automatic in its operation once a process is established.

4. Existing calculating machinery is capable of calculating $\phi(x)$ as a function of x by steps. Thus, if x is defined in the interval $a<x<b$ and $\phi(x)$ is obtained from x by a series of arithmetical operations, the existing procedure is to compute step (1) for all values of x in the interval $a<x<b$. Then step (2) is accomplished for all values of the result of step (1), and so on until $\phi(x)$ is reached. This process, however, is the reverse of that required in many mathematical operations. Calculating machinery designed for application to the mathematical sciences should be capable of computing lines instead of columns, for very often, as in the numerical solution of a differential equation, the computation of the second value in the computed table of a function depends on the preceding value or values.

Fundamentally, these four features are all that are required to convert existing punched card calculating machines such as those manufactured by the International Business Machine Company into machines specially adapted to scientific purposes. Because of the greater complexity of scientific problems as compared to accounting problems, the number of arithmetical elements involved would have to be greatly increased.

IV. The Mathematical Operations which should be Included

The mathematical operations which should be included in an automatic calculating machine are:

1. The fundamental operations of arithmetic

   a. addition

   b. subtraction

   c. multiplication

   d. division

2. Positive and negative numbers

3. Parentheses and brackets

   a. $(\ ) + (\ )$

   b. $[(\ ) + (\ )] \cdot [(\ ) + (\ )]$

   c. Etc.

4. Powers of numbers

   a. Integral

   b. Fractional

5. Logarithms

   a. Base 10

   b. All other bases by multiplication

6. Antilogarithms or exponential functions

   a. Base 10

   b. Other bases

7. Trigonometric functions

8. Anti-trigonometric functions

9. Hyperbolic functions

10. Anti-hyperbolic functions

11. Superior transcendentals

   a. Probability integral

   b. Elliptic function

   c. Bessel function

With the aid of these functions the processes to be carried out should be:

12. Evaluation of formulae and tabulation of results

13. Computation of series

   a. Finite

   b. Infinite

14. Determination of the real roots of equations

15. Solution of ordinary differential equations of the first and second order

16. Numerical integration of empirical data

17. Numerical differentiation of empirical data

V. The Mathematical Means of Accomplishing the Operations.

The purpose of this section is to describe the mathematical processes which may be made the basis of design of an automatic calculating machine. In the case of every operation considered it should be noted that the formulae suggested reduce the operation to a repetitive sequence.

1. The fundamental arithmetical operations require no comment, as they are already available, save that all the other operations must eventually be reduced to these in order that a mechanical device may be utilized.

2. Fortunately the algebra of positive and negative signs is extremely simple. In any case only two possibilities are offered. Later on it will be shown that these signs may be treated as numbers for the purposes of mechanical calculation.

The use of parentheses and brackets in writing a formula requires that the computation must proceed piecewise. Thus, a portion of the result is obtained and must be held pending the determination of some other portion, and so on. This means that a calculating machine must be equipped with means of temporarily storing numbers until they are required for further use. Such means are available in counters.

4. Integral powers of numbers may be obtained by successive multiplication, and fractional powers by the method of iteration. Thus, if it is required to find $5^{1/3}$,

$$y = f(x) = x^3 - 5 \tag{a}$$

and

$$x_n = x_{n-1} - \frac{f(x_{n-1})}{f'(x_{n-1})} \tag{b}$$

$$x_n = x_{n-1} - \frac{x_{n-1}^3 - 5}{3x_{n-1}^2} \tag{c}$$

or

$$x_n = \frac{2}{3} x_{n-1} + \frac{5}{3x_{n-1}^2}$$ (d)

Let

$$x_0 = 2$$

$$x_1 = \frac{4}{3} + \frac{5}{12} = \frac{21}{12}$$ (e)

$$x_2 = \frac{42}{36} + \frac{5 \times 144}{3 \times 441} = 1.166 + 0.544$$

$$= 1.710$$

which is the cube root of 5 to four significant figures. In general the rth root of θ is given by the iteration of the expression:

$$x_n = \left(1 - \frac{1}{r}\right) x_{n-1} + \frac{\theta}{r x_{n-1}^{r-1}}$$ (f)

Finally, if r is not an integer recourse may be had to the mechanical table of logarithms later to be described.

5. To supply a mechanical device with a complete mathematical function over a wide range of values would require an impossible amount of apparatus. To avoid this difficulty several artifices may be employed. In the case of logarithms, let it be required to find

$$y = \log_{10} x$$ (a)

Then

$$x = 10^y$$

$$= 10^{a + \frac{b}{10} + \frac{c}{100} + \frac{d}{1000} + \cdots}$$ (b)

where $a, b, c, \ldots$ are all integers. If x is restricted to have values no larger than 10^{10}, then

$$0 \le a, b, c, .. \le 9$$ (c)

Equation (b) may then be written

$$x = 10^a \times (10^{\frac{1}{10}})^b \times (10^{\frac{1}{100}})^c \times (10^{\frac{1}{1000}})^d \times \cdots$$ (d)

We may now form a table consisting of 100 numbers:

	0	1	2	3	4	5
10	1	10.000	100.00	1000.0	...	
$10^{1/10}$	1	1.2589	1.585	1.995	...	
$10^{1/100}$	1	1.0238	1.0471	1.0715	...	
$10^{1/1000}$	1	1.0023	1.0046	1.006	...	
$10^{1/10000}$	.	...	...	...	...	

giving the integral powers from 0 to 9 inclusive of 10, $10^{1/10}$, $10^{1/100}$, etc. Then, if it is required to find $\log_{10}2104$, for instance, choose the largest number in the first row which, when divided into 2104, still leaves a result greater than unity. Thus,

$$\frac{2.104}{1000} = 2.104 - - - - 3$$

where 1000 was taken from the 3rd column. Continuing,

$$\frac{2.104}{1.995} = 1.054 - - - - 3$$

$$\frac{1.054}{1.0471} = 1.006 - - - - 2$$

$$\frac{1.006}{1.006} = 1.000 - - - - 3$$

Hence,

$$\log_{10}2104 = 3.323; \tag{e}$$

this is correct to the last figure.

Thus it is seen that the computation of 10 significant figure logarithms may be reduced to ten discriminations, each in a field of ten, and eight divisions; eight because the first consists of moving the decimal point, a process as effortless in mechanical as in mental computation, and the last division need not be carried out.

6. The process of finding anti-logarithms may be reduced to a reversal of the logarithmic process. Thus, if

$$y = 10^x \tag{a}$$

then

$$y = 10^{a + \frac{b}{10} + \frac{c}{100} + \cdots} \tag{b}$$

$$= (10)^a \times (10^{\frac{1}{10}})^b \times (10^{\frac{1}{100}})^c \cdots \tag{c}$$

and repetitive discrimination and multiplication suffices.

7. The trigonometric functions most commonly used are the sine and cosine, and from these all other trigonometric functions may be computed easily. Either of these functions may be computed from the other, but in the expansion of Fourier series both sines and cosines are required. Therefore, it seems worth while to consider mechanical means of computing both the functions.

On expanding $\sin(a + h)$ by MacLauren's Theorem,

$$\sin(a+h) = \sin a + \frac{\cos a}{1} h - \frac{\sin a}{2} h^2 - \frac{\cos a}{6} h^3 + \frac{\sin a}{24} h^4 - \cdots \tag{a}$$

If now,

$$\theta = a + h \tag{b}$$

and

$$-\pi/2 \le \theta \le \pi/2 \tag{c}$$

twenty values of a may be chosen, as

$$a = \pi/2, \, 9\pi/20, \, 4\pi/5, \, \ldots -9\pi/20 \tag{d}$$

Then the maximum value of h is

$$h = \pi/20 = 0.15729\ldots \tag{e}$$

and ten terms of the series suffice for determining $\sin\theta$ significant figures, at most. On the average approximately five terms are sufficient. The process of computing sines is thus reduced to discriminations of one number in a field of twenty, and the computation of a series of at most ten terms.

The process for computing the cosine is exactly the same, and from these all other trigonometric functions may be determined arithmetically by

$$\csc\theta = 1/\sin\theta \tag{f}$$

$$\sec\theta = 1/\cos\theta \tag{g}$$

$$\tan\theta = \sin\theta/\cos\theta \tag{h}$$

Thus a field of 200 numbers of sufficient to supply all trigonometric functions.

8. The inverse trigonometric functions may also be determined by MacLauren's Theorem, but since $\sin^{-1}\theta$ and $\tan^{-1}\theta$ occur more often than any other inverse trigonometric function, these should be selected and any others computed from them.

9. Similar methods might be applied to the computation of the hyperbolic functions, but it is questionable if special apparatus should be initially installed for their determination since the hyperbolic functions may all be defined in terms of exponentials computable from the logarithmic device already suggested.

10. Similar comments apply to the inverse hyperbolic functions.

11. A great many functions may be similarly treated, and if the design of the automatic calculating machine proceeds so that a given device can be changed from one function to another rapidly, all such functions may be included in the scope of the machine. Means of accomplishing this will be suggested later.

12. Given a suitable supply of transcendental functions, the evaluation of formulae is reduced to arithmetic. If a formula is to be evaluated for a wide range of the independent variable, the process becomes repetitive. Means for accomplishing this will be discussed later.

13. The computation of closed series such as

$$y = a_0 + a_1 x + a_2 x^2 + a_3 x^3 \tag{a}$$

is most easily accomplished by the sequence:

$$a_3$$

$$a_3 x$$

$$a_3 x + a_2$$

$$a_3 x^2 + a_2 x \tag{b}$$

$a_3 x^2 + a_2 x + a_1$

$a_3 x^3 + a_2 x^2 + a_1 x$

$a_3 x^3 + a_2 x^2 + a_1 x + a_0 = y$

In the case of infinite series the computation may be reduced to successive multiplications and additions. Thus, if

$$y = a_0 + a_1 x + a_2 x^2 + a_3 x^3 \ldots \qquad (c)$$

$$= a_0$$

$$+ \frac{a_1}{a_0} x \cdot a_0$$

$$+ \frac{a_2}{a_1} x \cdot a_0$$

$$+ \frac{a_3}{a_2} x \cdot \frac{a_2}{a_1} x \cdot \frac{a_1}{a_0} x \cdot a_0$$

$$+ \frac{a_4}{a_3} x \cdot \frac{a_3}{a_2} x \cdot \frac{a_2}{a_1} x \cdot \frac{a_1}{a_0} x \cdot a_0 \qquad (d)$$

$$+ \ldots$$

and

$$y = A_0$$

$$+ A_1 x \cdot A_0$$

$$+ A_2 x \cdot A_1 x \cdot A_0$$

$$+ A_3 x \cdot A_2 x \cdot A_0$$

$$+ A_4 x \cdot A_3 x \cdot A_2 x \cdot A_1 x \cdot A_0 \qquad (e)$$

where

$$A_0 = a_0; \; A_1 = a_1/a_0; \; A_2 = a_2/a_1; \; \ldots \qquad (f)$$

Thus each term of the series is obtained from the last through multiplication by a coefficient and the value of the independent variable.

14. Any mechanical device that can evaluate formulae can also determine the real roots of algebraic and transcendental equations provided only that in the evaluation of the formulae the successive values of the independent variable are the successive values of the dependent variable computed; thus, consider

$$x + \log_{10} x = 1/2 \tag{a}$$

given that x is in the neighborhood of 1/2. A succession of eleven approximations suffices to give

$$x = 0.672384 \tag{b}$$

Or, let the equation be the famous cubic of Wallis,

$$x^3 - 2x - 5 = 0 \tag{c}$$

the iterative equation is

$$x_n = x_{n-1} - \frac{x_{n-1}^3 - 2x_{n-1} - 5}{3x_{n-1}^2 - 2} \tag{d}$$

Three approximations give

$$x = 2.09455148... \tag{e}$$

The root of this equation has been computed to 150 significant figures. Note that again the process is purely repetitive after being started.

15. The solution of ordinary differential equations of any order can usually be accomplished to any degree of accuracy by expansion into infinite series by MacLauren's Theorem for any specified boundary demands. Under certain circumstances the series may be rapidly convergent and the method offers excellent means for numerical solution.

However, when the equation has complicated functions of x as coefficients of the various derivatives of y, and the independent variable itself occurs in complicated functions, the various successive derivatives necessary to the series expansion may involve a prohibitive amount of labor. For such cases various methods of numerical solution have been devised, such as those of Adams, Runge-Kutta, and others.

Of these, the method of Runge-Kutta is probably best adapted to mechanical computation because the method of solution depends entirely on the evaluation of a repetitive sequence. Thus, if

$$\frac{dy}{dx} = f(x,y) \tag{a}$$

and

$$K_1 = f(x_0, y_0)\Delta x$$

$$K_2 = f\left(x_0 + \frac{\Delta x}{2}, y_0 + \frac{K_1}{2}\right)\Delta x \qquad \text{(b)}$$

$$K_3 = f\left(x_0 + \frac{\Delta x}{2}, y_0 + \frac{K_2}{2}\right)\Delta x$$

$$K_4 = f(x_0 + \Delta x, y_0 + K_3)\, \Delta x$$

Then,

$$\Delta y = (x_0 + \Delta x, y_0 + K_3)\, \Delta x$$

and

$$y_1 = y_0 + \Delta y$$

$$x_1 = x_0 + \Delta x \qquad \text{(c)}$$

The process may now be repeated to find x_2, y_2, and so on. The inherent error of this process is of the order of Δx^5; hence, if Δx is taken as 0.1, the solution will be correct to the fourth place of decimals, and doubtful in the fifth.

The method can be applied to simultaneous equations of the first order, and hence to second order equations.

Since the method involves nothing other than the evaluation of formulae, a mechanical device suitable for such evaluation is prepared to perform this type of numerical integration.

16. The numerical integration of empirical data may be carried out by the rules of Simpson, Weddle, Gauss, and others. All these rules involve sums of successive values of y multiplied by specified numerical coefficients. Hence the only new mechanical component involved in a means of mechanically introducing a list of numbers. Means of accomplishing this will be discussed later.

17. Numerical differentiation of empirical data is best accomplished by means of a difference formula. Most experimental observations are of such an accuracy that fifth differences may be neglected by taking observations sufficiently close together. If, then, all differences above the fifth may be neglected, the process of numerical differentiation may be carried out by a fifth difference engine such as originally designed by Babbage. Such a device can, however, be assembled from standard addition-subtraction machines with but a few changes. The

differentiating apparatus would also be applicable to many other problems. In fact, most of the problems already discussed may under certain circumstances be solved by application of difference formulae.

VI. Mechanical Considerations

In the last section it was shown that even complicated mathematical operations may be reduced to a repetitive process involving the fundmanetal rules of arithmetic. At the present time the calculating machine of the International Business Machines Company are capable of carrying out such operations as:

$$A + B = F$$

$$A - B = F$$

$$AB + C = F$$

$$AB + C + D = F \tag{a}$$

$$A + B + C = F$$

$$A - B - C = F$$

$$A + B - C = F$$

In these equations, A, B, C, D are tabulations of numbers on punched cards, and F, the result, is also obtained through punched cards. The F cards may then be put through another machine and printed or utilized as A, B, . . . cards in another computation.

Changing a given machine from any of the operations (a) to any other is accomplished by means of electrical wiring on a plug board. In the hands of a skilled operator such changes can be made in a few minutes.

No further effort will be made here to describe the mechanism of the International Business Machines. Suffice it to say that all the operations described in the last section can be accomplished by these existing machines when equipped with suitable controls, and assembled in sufficient number. The whole problem of design of an automatic calculating machine suitable for mathematical operations is thus reduced to a problem of suitable control design, and even this problem has been solved for simple arithmetical operations.

The main features of the specialized controls are machine switching and replacement of the punched cards by continuous perforated tapes.

In order that the switching sequence can be changed quickly to any possible sequence the switching mechanism should itself utilize a paper tape control in which mathematical formulae may be represented by suitably disposed perforations.

VII. Present Conception of the Apparatus

At present the automatic calculator is visualized as a switchboard on which are mounted various pieces of calculating machine apparatus. Each panel of the switchboard is given over to definite mathematical operations.

1. International Business Machines utilize two electric potentials, 120 a.c. for motor operation, and 32 volts d.c. for relay operation, etc. A main power supply panel would have to be provided including control for a 110 volt, a.c./32 volt d.c. motor generator and adequate fuse protection for all circuits.

2. Master Control Panel: The purpose of this control is to route the flow of numbers through the machines and to start operation. The processes involved are

   a. Deliver the number in position (x) to position (y)

   b. Start the operation for which position (y) is intended.

The master control must itself be subject to interlocking to prevent the attempt to remove a number before its value is determined, or to begin a second operation in position (y) before a previous operation is finished.

It would be desirable to have four such master controls, each capable of controlling the entire machine or any of its parts. Thus, for complicated problems the entire resources could be thrown together; for simple problems, fewer resources are required and several problems could be in progress at the same time.

3. The progress of the independent variable in any calculation would go forward by equal steps subject to manual readjustment for change in the increment. The easiest way to obtain such an arithmetical sequence is to supply a first value, x_0, to an adding machine, together

with the increment Δx. Then successive additions of Δx will give the sequence desired.

There should be four such independent variable devices in order to

   a. Calculate formulae involving four variables,

   b. Operate four master controls independently.

4. Certain constants: many mathematical formulae involve certain constants such as ε, π, $\log_{10} \varepsilon$, and so forth. These constants should be permanently installed and available at all times.

5. Mathematical formulae nearly always involve constant quantities. In the computation of a formula as a function of an independent variable these constants are used over and over again. Hence the machine should be supplied with 24 adjustable number positions for these constants.

6. In the evaluation of infinite series the number 24 might be greatly exceeded. To take care of this case it should be possible to introduce specific values by means of a perforated tape, the successive values being supplied by moving the tape ahead one position. Two such devices should be supplied.

7. The introduction of empirical data for non-repetitive operations can be accomplished best by standard punched card magazine feed. One such device should be supplied.

8. At various stages of a computation involving parentheses and brackets it may be necessary to hold a part of the result pending the computation of some other part. If results are held in the calculating unit these elements are not available for carrying out succeeding steps. Therefore it is necessary that numbers may be removed from the calculating units and temporarily stored in storage positions. Twelve such positions should be available.

9. The fundamental operations of arithmetic may be carried on three machines.

   a. Addition and subtraction

   c. Multiplication

   d. Division.

Four units of each should be supplied in addition to those directly associated with the transcendental functions.
   The permanently installed mathematical functions should include

10. Logarithms.

11. Anti-logarithms

12. Sines

13. Cosines.

14. Inverse Sines.

15. Inverse tangents.

16. Two units for MacLauren Series expansion of other functions as needed.

17. In order to carry out the process of differentiation and integration on empirical data, adding and subtracting accumulators should be provided sufficient to compute out to fifth differences.

18. All results should be printed, punched in paper tapes, or in cards at will. Final results would be printed. Intermediate results would be punched in preparation for further calculations.

   The above is a rough outline of the apparatus required, and it is believed that this apparatus, controlled by automatic switching, would care for most of the problems encountered.

VIII. Probable Speed of Computation

An idea of the speed attained by the International Business Machines can be had from the following tabulation of multiplication in which 2×8 refers to the multiplication of an 8 significant figure number by a 2 significant figure number, zeros not counted.

	Products per hour
2×8	1500
3×8	1285
4×8	1125
5×8	1000
6×8	900
7×8	818
8×8	750

In the computation of 10 place logarithms the average speed would be about 90 per hour. If all the 10 place logarithms of the natural numbers from 1000 to 100,000 were required, the time of computation would be approximately 1100 hours, or 50 days, allowing no time for addition or printing. This is justified since these operations are extremely rapid and can be carried out during the multiplying time.

IX. Suggested Accuracy

Ten significant figures has been used in the above examples. If all numbers were to be given to this accuracy it would be necessary to provide 23 number positions on most of the computing components, 10 to the left of the decimal point, 12 to the right, and one for plus and minus. Of the twelve to the right two would be guard places and thrown away.

X. Ease of Publication of Results

As already mentioned, all computed results would be printed in tabular form. By means of photo-lithography these results could be printed directly without type setting or proof reading. Not only does this indicate a great saving in the publishing of mathematical functions, but it also eliminates many possibilities of error.

Aiken's First Machine: The IBM ASCC/Harvard Mark I

Robert Campbell

Completing the Machine at Endicott

First Meeting

My first meeting with Howard Aiken—then Lieutenant Commander Howard Aiken—was in New York, during the Christmas season of 1941, some three weeks after Pearl Harbor. He had telephoned me from Boston to set up a meeting during his brief stopover in New York, while he was en route to Yorktown, Virginia. We met in Grand Central Station. Aiken's fiancée, Agnes Montgomery, was traveling with him. We had a discussion lasting perhaps two hours. Aiken explained that an automatic calculating machine was under construction at the IBM engineering laboratories in Endicott, New York. The machine, he said, was the result of his initial conception, and had been designed jointly by him and IBM engineers. After his initial design participation, he had been called up to active duty in the Naval Reserve, and was teaching in a course that he had structured at the Naval Mine Warfare School at Yorktown. (At that time "degaussing" equipment was being brought into use to counter the threat of magnetically detonated mines.) He needed a deputy to continue the technical liaison between Harvard and IBM. He described the assignment in terms of assisting with aspects of the final design, helping with problems, testing the machine, and paving the way for moving the completed machine to Harvard and establishing an operational computation facility.

During our meeting, Aiken gave me an overview of the machine and explained who the program's principal participants and decision makers were. He emphasized that Harvard Professors Chaffee, Mimno, and Shapley would have to be kept informed of plans and progress. He also indicated that Robert L. Hawkins, a machinist in the Cruft Laboratory shop at Harvard, would be sent on full-time assignment to

Endicott to learn how to operate and service the machine. Aiken was a very persuasive salesman, and the assignment sounded intriguing; I agreed to start work after talking with the people that he had mentioned.

Aiken expressed the hope that when the calculator was finished it could somehow be put to work for the Navy. I was in no position, however, to judge whether Navy use would really be feasible, or whether this was something that Aiken could successfully promote. In the end, his combination of technical and entrepreneurial skills, together with the demands of the war effort, led to the establishment of a Navy computer center at Harvard in an amazingly short time. But first the final design, construction, and testing had to be completed, and this took all of the next two years.

My Background

I had had experience in making statistical calculations by machine, and had a B.S. from MIT and an M.A. from Columbia in physics. I was enrolled in the doctoral program in physics at Harvard, where I was appointed a teaching fellow in physics for the 1941–42 academic year. During the fall of 1941 I had passed my qualifying examination for the Ph.D., and I expected to carry out my thesis work under Professor John H. Van Vleck. In addition to taking a broad range of courses in physics and applied mathematics, I had taught laboratory sections. For two years I had been a member of a research project at the Columbia-Presbyterian Medical Center, where I had carried out analyses of variance, correlations, and other statistical processes using rotary mechanical calculators, such as those made by Friden, Marchant, and Monroe. Perhaps it was this last experience that caused Aiken, while in Cambridge on Christmas leave from his post in Yorktown, to choose me as a likely candidate for the liaison assignment with IBM.

Persons Involved

At Harvard, the computer project was under the overall supervision of Emory L. Chaffee, Professor of Physics and Director of the Cruft Laboratory. Professor Harry R. Mimno also had a major role in the project. These were the people at Harvard with whom I worked most closely during the liaison period. I also had to keep in touch with Professor Harlow Shapley, Director of the Harvard College Observatory, who was one of the early supporters of Aiken at Harvard. I remember my meetings with him at the observatory with special interest. His knowledge and activities were very broad, and he had a desk of unique construction: it was octagonal in shape and built on a swivel. When he "changed hats," he would correspondingly rotate the desk.

At IBM the design and construction of the calculator were under the overall supervision of Clair D. Lake. Under him were Frank E. Hamilton and Benjamin M. Durfee, who were responsible for the detailed design. While in Endicott part-time during 1942, 1943, and early 1944, I had my primary contacts with these engineers and with Bob Hawkins. Bob, while at Endicott, became fully familiar with the design and operation of the machine, and assisted with its checkout and testing. After the machine was moved to Harvard, he had a key roll in the operation and maintenance of the machine for more than 10 years.

Previous Work on the Calculator
When I joined the project, in January of 1942, the design of the machine had largely been completed, and a considerable portion of it had been assembled. Aiken had worked closely with the IBM engineers to determine exactly how to apply IBM devices and design approaches to achieve the required functions. Most of the necessary building blocks were already available in the IBM technical inventory, although the chosen decimal counter and the chosen relay had not yet been utilized in commercial equipment. To provide a means of introducing instructions, however, Aiken needed two devices not then available from IBM: a tape punch and a tape reader. (He did not believe that punched cards would be suitable for this purpose.) After some discussion, it was decided to design units for perforating and reading tape made from uncut card stock of the same width as standard IBM cards. These units were developed using standard design approaches to meet Aiken's functional requirements. It was also decided to utilize this type of perforated tape for introducing lists of numerical values, including tables of functions.

The machine that emerged was entirely electromechanical. It utilized rotary decimal counters, driven mechanically but controlled electrically, for internal storage, addition, and subtraction. It had a multiply-divide unit and special built-in procedures for computing logarithms, exponentials, and sines and for carrying out interpolation. Numerical inputs were made via punched cards, function ("value") tapes, and manually set groups of dial switches. Numerical outputs were printed using solenoid-actuated typewriters; for reentrant data, outputs were punched into cards. (More details of the machine's design are given later in this chapter.)

Aiken spent two summers, and other time as available, working with the IBM engineers. He participated directly in much of the detailed design of the machine. The first detailed circuit drawings are dated 1939. By the end of 1940, test assemblies of machine units had begun. Then, on 21 April 1941, Aiken was called up for active duty in the Naval Reserve. He was no longer able to continue his participation in the work at Endicott.

My Work at Endicott

When I first went to Endicott, at the beginning of 1942, a few portions of the machine still had to be designed, and some earlier designs required modification. The only design record from this period that I still have is a memorandum, dated 9 January 1942, written by Hamilton. This documents the results of a discussion involving me, Hamilton, and Durfee and concerning some changes that needed to be made in the built-in procedures for calculating logarithms and exponentials. The changes were modifications to reports of Aiken's dated 6 and 11 July 1939.

My general liaison work at Endicott involved reviewing circuit diagrams and timing charts, supplying some procedural and design details for the built-in functional procedures, proposing how some units and functions could be tested, and assisting with the testing. I also computed some of the many constants that needed to be wired into the machine. Later I formulated and programmed five simple illustrative problems for test and demonstration purposes.

During 1942, 1943, and early 1944 I was spending, on average, 20–25 percent of my working time at Endicott. The remainder was at first spent in continuing my physics courses and my teaching in the physics laboratory. Before long, however, I began spending all my time at Harvard teaching in the "Pre-Radar" instruction course—an intensive three-month course in the fundamentals of radio engineering that was given to Army, Navy, and Marine officers to prepare them for the Radar program at MIT. I taught "quiz sections," oversaw experiments in the laboratory, and lectured on electric filters and (once) detection. The Pre-Radar course was under the overall management of Professor Chaffee, although the detailed operating management was under Professor Russell Tatum. Luckily, I was able to adjust my teaching schedule to permit me to spend the necessary time at IBM.

I had only limited contact with Aiken during this period. He was completely involved in his teaching at the Naval Mine Warfare School, together with, I believe, other special Naval assignments. Hamilton and I made a trip from Endicott to Yorktown to see Aiken in February of 1942, and I also made one or two trips on my own. During these visits we reviewed progress and plans, and discussed any open technical issues.

Relative Contributions of Aiken and IBM

The Automatic Sequence Controlled Calculator was the result of combining Aiken's concept with IBM's components and design approach. Without Aiken there would have been no machine. He understood the need for better and more automatic means of computation. He was familiar with many of the types of problems that needed to be solved, and with the numerical methods that would be required. He also understood the repertoire of basic functions that the machine would have to perform, and in particular that means must be provided for automatic sequence (or program) control. He was aware that partial results would need to be stored automatically so that they could be retrieved and utilized as required in subsequent computational steps. He fully saw how IBM technology could be used to implement his concept, and he contributed materially to the detailed design. In addition to his technical abilities, he was an effective salesman, entrepreneur, and leader.

IBM was a full partner in the machine's development. IBM already had most of the key components and design techniques necessary to realize an electromechanical version of Aiken's concept. IBM's technology was based upon a mechanical drive, electrical controls, and electrical communication between units. (The other major US maker of punched cards, Remington Rand, utilized mechanical communication between the card feed and the accumulators in its tabulator.) Few if any other American companies had as complete and as flexible a set of computation and data processing building blocks, except perhaps for the Bell Telephone Laboratories. IBM's punched-card technology was a good springboard for the realization of the machine, although it did limit the machine in some respects—especially in basic computing speed. The major new devices coming out of the Aiken-IBM effort were the tape perforating and reading units needed in connection with the sequence (instruction) and numeric tapes, and even these were modifications of existing IBM design concepts. Thus, IBM supplied most of the necessary technology, contributed the major part of the engineering and design effort, and built the machine. IBM funded the entire effort and gave the completed machine to Harvard. The patent that IBM took out on the machine lists Aiken, Lake, Hamilton, and Durfee as the inventors.

Description of the Calculator

Overall Characteristics and Functions

Mark I was a pioneering effort, a highly capable machine for the era in which it was conceived and designed. However, by today's standards the machine seems limited in its functions, awkward to program and operate, and extremely slow. The internal functions of the machine were built around 72 storage and accumulating registers, each register having 24 decimal counters, so as to handle a numerical "word" containing 23 decimal digits plus an algebraic sign (figure 1). A word transferred into any register was automatically added to the word previously stored in that register. Negative numbers were represented by the nines complement. To effect a subtraction, the machine would add the nines complement of the subtrahend to the augend. The machine also had a multiply-divide unit and built-in controls for calculating three elementary functions: logarithms, exponentials, and sines. Curiously, there was no built-in square root function.

The reason for the 23-decimal-digit precision is not generally known. I once queried Aiken on this point, and he replied that it had been planned to utilize the calculator to recompute the orbits of all the major planets in the solar system, and that a precision of 23 digits was judged to be necessary for this task. As it turned out, however, the machine was never used for this purpose.

Numerical words were introduced into the machine from either of two punched-card readers (usually called "feeds") or from manually set decimal dial switches. There were 60 groups of dial switches, each group having 24 switches to represent one numeric word. In addition, lists of numeric words or tables of functions could be introduced from punched "value tapes" using any of three tape readers ("interpolator units"). A table of functions was represented on the tape by a set of data blocks. Each block contained an argument X and a related set of interpolation coefficients. The machine had built-in controls to search the tape for the appropriate argument, and then to utilize the information in the block to perform an nth-degree polynomial calculation. Numerical outputs could be obtained in printed form by using either

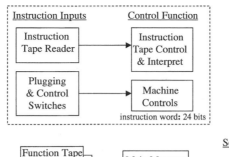

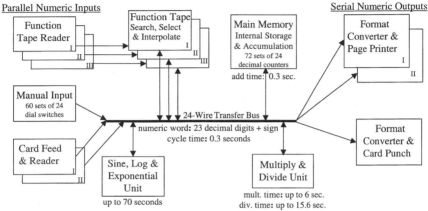

Figure 1

A block diagram showing the main functional units of Mark I, emphasizing the flow of numerical data from inputs (card readers and manually set dial switches), through and between storage and processing units, to outputs (page printers and a card punch). There are also special input units that position and read numerical punched tapes. Numeric transfer among all these units is via a 24-wire bus (indicated by a heavy line), corresponding to the machine's numeric "word" of 23 decimal digits plus algebraic sign. At the top is a special punched-tape reader for instructions, supplemented by plugboards for additional control functions. Key capacity parameters and operating times are also indicated (The basic source of information on Mark I is the Manual of Operation, originally published in 1946. Though exhaustively complete in describing the design and operation of Mark I, the manual does not have any overall block diagram. Hence, this figure has been prepared to illustrate the main points made in the text.) We are indebted to Gary M. C. Bean for implementing the graphic design.

of two solenoid-actuated typewriters, or as punched cards by using a standard IBM "summary punch."

Instructions were introduced into the machine by means of punched sequence tapes fed into a tape reader. Only one such reader was provided in the original design. Because the sequence-tape format was different from the value-tape format, the two types of tape readers, though similar, were not identical. An endless tape loop could be used to carry out a computing routine any desired number of times. In the original machine, program branching was accomplished manually; the operator could move the tape ahead or back, or change tapes, upon receipt of a suitable signal from the calculator. In any given computer run, the decimal point in a numeric word could be set in any desired position. That position was established by means of plugboards that controlled the columnar shifts necessary for multiplication, division, the three built-in functions, and interpolation. A plugboard was also used to control the printing of output numbers on each typewriter.

The instruction word, a 24-digit binary number, was represented by one row of holes on the sequence tape. All programming was done in machine language, using explicit addresses.

The machine had a basic cycle time of 0.3 second. This was the time necessary to transfer a numeric word from one storage register to another and to add the transferred word to, or subtract it from, the number already in the second register. The same time was required to read from a group of dial switches or from a card feed into a storage register. Multiplication took up to 6 seconds, and division up to 15.6 seconds, depending upon the number of digits utilized in the multiplier or quotient.

In addition to being used on line to prepare output cards, the card punch could be used off line to prepare numeric input cards manually. The cards utilized had the standard IBM format. Value tapes (for the interpolator units) and sequence tapes were punched manually using specially designed keyboard-to-tape equipment, with different keyboards for the two tape types.

Layout of the Calculator

The calculator was built as a line of panel-like structures about 8 feet high and 51 feet long, with two panels extending for about 6 feet at right angles to, and behind, the main line of units. Looking at the machine from the front, one saw first, from left to right, two panels, each with 30 groups of dial switches, for manually entered numerical inputs. Next were six panels, each with 12 storage registers, then the multiply-divide unit and the functional and interpolation control units. To the right of these were the four tape units—three for value or function tapes and one for sequence or instruction tapes. At the extreme right, two output typewriters sat on a shelf, with two input-card readers and the output-card punch below.

The main driveshaft ran most of the length of the front panels, coupling the storage registers, the registers in the multiply-divide, functional, and interpolation control units, the four tape readers, and the two card readers. The relays used to control readout of the dial switches, and readin and readout of the storage registers and other special registers, were on the back of the front panel. The panels extending back at right angles contained most of the other control relays. The various plugboards were near the center of the front-panel array. The entire front-panel array, and the back-extending panels, were enclosed in a decorative glass case with sliding glass doors. The case also had the practical function of providing some protection from dust. The main driveshaft was the basic source of synchronization throughout the machine. A 24-wire bus, extending the length of the machine and having many branches, provided an electrical path for all transfers of numeric words between major units.

Major Components

As I have mentioned, almost all of the machine's major components were already available at IBM, either as elements of standard commercial products or as devices that had been developed and were ready for possible use in production equipment. Even the principal exceptions to this general rule—the devices utilized to handle, punch, and read the value and sequence tapes—made use of standard uncut IBM punched-card stock.

The storage and accumulating registers that were used throughout the machine were the means for all internal storage of numeric words, as well as all addition and subtraction, and hence were the basis of all computation. They were composed of sets of ten-position rotary counters, having one position for each of the digits from 0 through 9. All counters throughout the machine were driven by an interconnected system of driveshafts. A counter had the digit that it represented changed by n units by being clutched into its driveshaft for n units of time. This was accomplished by supplying an electrical impulse to the counter's electromagnet n units of time before all the engaged counters were mechanically disengaged. Each counter had an electrical readout that was used when the digit standing in the counter was to be transferred to another counter. A counter had one external input connection for each of the digits 0 through 9, plus an eleventh input connection for carry. It had an output common connection, plus another output for carry. Through a pair of brushes the position of the counter selected one of the 0 through 9 input leads. To read out a nines complement, the machine used control relays to reverse the "meaning" of the 0 through 9 external leads so that they represented the digits 9 through 0 respectively. The characteristics of the decimal counters were the basic determinants of the 0.3-second transfer-and-add cycle time of the machine.

The transfer of numeric words between registers, and the combining of these transfers to carry out the more complex processes of multiplication, division, functional calculation, and interpolation, were all under the control of relays. These were 4-, 6-, or 12-position double-throw devices. In all types, a solenoid, pulling against a spring return, actuated an armature that controlled the position of the contacts, either in the normally closed position (solenoid not energized) or in the normally open position (solenoid energized). The electrical impulses used to energize the control relays, as well as those used to clutch in the counters, were generated by cam-operated contactors, all driven from the synchronized shaft system. All circuits were made and broken by these contactors, not by the relay contacts. Some relays also had a second solenoid (called a "hold coil") that would keep the relay in the normally open position after the original source of current had been removed.

Numeric words were transferred between registers by electrical impulses traveling through a 24-wire bus, one wire for each decimal digit. To transfer a word from register A to register B, the machine first actuated relays to connect the readout circuits of the counters of register A to the bus, and also to connect the input solenoids of the counters of register B to the bus. Then electrical impulses generated by the appropriate cam-operated contactors would traverse the circuits that had been established. The bus was used for numeric word transfers involving the 72 storage registers, the various input and output registers used for multiplication, division, functional calculations, and interpolation. It also transferred input data from the 60 groups of dial switches and from the two card feeds, as well as output data to the registers used for temporarily storing data to be printed on either typewriter, or to be utilized by the card punch. Since there was only one bus in the calculator, only one numeric word could be transferred between any pairs of the units listed above at one time. While processes not using the bus were being carried out—for example, the steps internal to a multiplication or division—the bus could be used for additions, subtractions, and other processes. The two input-card feeds were mechanically coupled to the main driveshaft. The 80 card columns (not necessarily all used) were read in parallel; within each column the reading was serial, being synchronized with the operation of the storage counters. The two output typewriters and the output-card punch handled a numeric word serially and operated at their own independent speeds, after the word to be printed or punched had been read from the bus into one of three buffer registers. The typewriters, the card feeds, and the card punch were standard IBM equipment, modified as necessary for incorporation into the system.

The punched "paper" tapes (actually they were uncut card stock) utilized to introduce instructions into the machine (we generally called them sequence tapes) represented one instruction by a single row of 24 hole positions. The tape handling and reading mechanism read one instruction and stepped to the next in a single machine cycle. Unlike the card readers, which sensed electrically, by using brushes, the tape readers sensed mechanically, by using an array of 24 pins.

The punched paper tapes containing numerical values required four rows of 24 hole positions each to represent a word of 23 digits plus an algebraic sign. The tape handling and reading mechanisms for these tapes read a complete numeric word and stepped to the next word in one machine cycle. Four rows of holes on a numeric tape occupied as much space as two rows of holes on the sequence tape. Hence the numeric tape mechanisms had pins to read 96 hole positions, and advanced twice as far as the sequence-tape mechanism in each step. When clutched in, both types of tape readers were driven directly from the main driveshaft.

Thus the representations of numbers and instructions were completely unrelated, and the two types of tape handling units were not interchangeable. The design of the machine did not allow either for instructions to be handled internally like numbers or for instructions to be modified by arithmetic operations.

Another major component of the machine was the plugboard. Plugboards were used to provide temporary connections—connections that could be changed manually between computer runs. One group of plugboards was related to the position of the decimal point. These provided the columnar shifts that were required when a product, quotient, logarithm, exponential, sine, or result of interpolation was read out from its specialized unit into the bus. Plugboards were also used to control the format in output printing, to shift words transferred from the card feeds or to the card punch, and for various other purposes.

Speed of Operation

In the basic cycle time of 0.3 second, the machine would act on the previously read instruction by transferring a numeric word via the bus and performing an addition or subtraction, would also read the next instruction from the sequence tape, and would step the tape to the instruction beyond. Reading a numeric word from a card feed or from a value tape also took one cycle. Each 0.3-second cycle was divided into 16 segments or "points." Nine of these were used for advancing a storage counter up to nine units, two were for handling the carry, and the remainder were used to position the control relays for the next cycle.

Table 1 shows the times necessary to complete major machine functions, assuming that computation was done to full precision and the decimal point lay between columns 15 and 16. The time for all operations except addition and subtraction was reduced if fewer significant figures were used in the calculation. Output printing was done at 3.4 characters per second, and output-card punching at about 2.6 characters per second.

Multiplication and Division

IBM proposed the approaches that were used for multiplication and division. Although IBM did not at the time provide a capability for division in its commercial equipment, it was able to piggy back on the design for multiplication. (J. W. Bryce of IBM's home office, who had been a key party in the early discussions with Aiken, entered the picture here.) If a product were to be generated by simple over and over addition of the multiplicand, under control of the digits of the multiplier, then a 23-digit multiplier would require, on average, more than 100 cycles (30 seconds). If division were to be performed similarly by over and over subtraction, it would take at least as long. In Mark I, however, these two processes were greatly speeded up by an approach in which the principal part of multiplication used one cycle for each two multiplier digits and the principal part of division used two cycles for each quotient digit.

Table 1

Function	Cycles	Seconds
Addition or subtraction	1	0.3
Multiplication	20	6.0
Division	38	11.4
Compute log$x$	228	68.4
Compute exp$x$	204	61.2
Compute sin$x$	199	60.0

In the multiplication process, the machine first computed a table of the first nine integral multiples of the multiplicand, then selected these two at a time according to successive pairs of digits in the multiplier in order to accumulate two partial products, and finally combined them to form the final product. To build up the multiples, the machine required three addition cycles and utilized six storage registers. Three of these were composed of special counters that could read out either the number stored or twice that number. The six registers were assigned to the multiples as follows: 1 and 2; 3 and 6; 4 and 8; 5; 7; and 9. Multiplication thus took one cycle to read in the multiplicand, three cycles to build up the nine multiples (and also read in the multiplier), one cycle for each nonzero pair of digits in the multiplier, and two final cycles to combine the two partial products and readout the final result. With full precision and no zero pairs, the time requirement was 20 cycles (6 seconds).

In division, the machine built up the first nine integral multiples of the divisor. The machine then took two cycles per quotient digit: one cycle to compare simultaneously all nine multiples with the dividend or remainder, and one cycle to subtract the chosen multiple. Division to obtain a full 23-digit quotient, having no zeros, took 52 cycles (15.6 seconds).

Built-In Functions

Although the methods used for performing multiplication and division were introduced by IBM, Aiken himself was responsible for the procedures for computing all other built-in functions and for the method of interpolation. Similar methods were used by the machine to calculate logarithms, exponentials, and sines. In each case a simple power series was used, after the range of the argument had been reduced by some preliminary selections and calculations. For a logarithm (exponential), the result of the series calculation was added to (multiplied by) numbers found by table lookups. The procedures could be used to calculate the functions to a precision of at least 21 digits if full precision were used in the multiplications and divisions involved.

The logarithm to the base 10 (log) of an argument X was determined as follows. First X was shifted so that its highest-order digit was in column 23; the magnitude of the shift determined the value of the characteristic of the logarithm. The shifted number, say X_0, was then converted by five successive divisions to the product of five factors, namely $X_0 = X_1 \cdot X_2 \cdot X_3 \cdot X_4 \cdot X_5$. Here X_1 was the first (highest-order) digit of X_0; X_2 was the first two digits of X_0/X_1; X_3 was the first three digits of $(X_0/X_1)/X_2$; X_4 was the first four digits of $((X_0/X_1)/X_2)/X_3$; and X_5 was the full result of $(((X_0/X_1)/X_2)/X_3)/X_4$. Here the Xs could take on the following values:

X_1 was one of the set 1 (1) 9

X_2 was one of the set 1.1 (0.1) 1.9

X_3 was one of the set 1.01 (0.01) 1.09

X_4 was one of the set 1.001 (0.001) 1.009

X_5 could be written as $1 + H$, where H was less than 0.001.

$\text{Log}X_0$ was then obtained as the sum of the logarithms of X_1, X_2, X_3, X_4, and X_5. The logarithms of X_1, X_2, X_3, and X_4 were obtained by built-in table lookups. The logarithm of $(1 + H)$ was computed by means of a Taylor-series expansion through the term in H^6, using six multiplications and five additions:

$$\log(1 + H) = (((((-C_6 \cdot H + C_5)H - C_4)H + C_3)H - C_2)H + C_1)H,$$

where the C_n were built-in constants having values equal to $((\log e)/n)$, e being the base of natural logarithms.

In determining the exponential to the base 10 (exp) of an argument X, considered to have two integral digits and up to 21 decimal digits, the machine dealt with X as the sum of six terms. Thus X was utilized in the form IJ.KLMF, where I, J, K, L, and M were the successive digits in X reading from the left, and F, which was less than 0.001, was X with its five highest-order digits changed to zeros. (For example, if X were the number 12.643759 . . . : I would be 1, J would be 2, K would

be 6, L would be 4, M would be 3, and F would be 0.000759 . . .)
Exp(X) was then computed as the product of the following:

$\exp(10 \cdot I)$

$\exp(J)$

$\exp(K/10)$

$\exp(L/100)$

$\exp(M/1000)$

$\exp(F)$.

The machine had permanently stored the 30 values of $\exp(K/10)$, $\exp(L/100)$, and $\exp(M/1000)$, and computed $\exp(F)$ by means of a power series. After the machine had multiplied together the exponentials relating to K, L, M, and F, the result was shifted by $10 \cdot I + J$ columns. The computation of $\exp(F)$ was based on the series

$$\exp(F) = 1 + C_1G + C_2G^2 + C_3G^3 + \cdots + C_6G^6,$$

where $G = F/\log(e)$, and the stored constants C_n were equal to $1/n!$.

In determining $\sin X$, the calculator first reduced X to the equivalent first quadrant angle X_0, while noting whether or not a minus sign needed to be affixed to the final result. Then, if X_0 was not greater than $\pi/4$, it used the sine series

$$\sin X = C_1X - C_3X^3 + C_5X^5 - C_7X^7 + \cdots + C_{21}X^{21}.$$

If, however, X_0 was greater than $\pi/4$, it computed $\cos Y$, where $Y = (\pi/2 - X_0)$, using the series

$$\cos Y = 1 - C_2Y^2 + C_4Y^4 - C_6Y^6 + \cdots + C_{20}X^{20}.$$

In the equations for $\sin X$ and $\cos Y$, C_n was equal to $1/n!$.

The procedures used for the built-in functions were designed to produce results to the full precision of the machine. If less than the full precision were required, the same sets of additions, subtractions, multiplications, and divisions still had to be carried out. But multipliers having fewer nonzero digits, and quotients calculated to fewer digits (through switch controls) would speed up the calculations to some extent. After acquiring some experience with the machine, we often found it advisable to use programmed procedures tailored to the specific precision requirements rather than the built-in procedures.

How Other Functions Were Obtained

Functions other than those just described could be obtained by a built-in interpolation procedure that utilized data recorded on a punched tape, and one of the three numeric tape handling mechanisms. The interpolation procedure was composed of two distinct parts: first, the positioning of the tape so that the desired numerical values could be read; and, second, reading the values off the tape and performing the interpolation calculation.

To introduce a function $F(X)$ into the calculator by this method, one first divided the range of interest of the argument X into N equal steps. For each step, a block of data was entered onto the tape. The block contained the identifying argument X_n, the functional value $F(X_n)$, and up to eleven associated interpolation coefficients. When the calculator began to determine $F(X)$, it first found the block of data for the tape argument X_n closest to X. Typically the functional tape would be endless; that is, it would be a tape loop. The machine read the argument where the tape was initially positioned, and automatically stepped to the desired tape argument by the shorter route. Once the tape had been positioned, the second part of the interpolation process could begin.

The built-in relay controls for interpolation were designed to handle one particular approach to the calculation. In this approach, the function $F(X)$ within the vicinity of a tape argument was represented by a Taylor series. The tape argument X_n was read from the tape together with the corresponding Taylor's coefficients, and the power series in $(X - X_n)$ was evaluated. Dial switches on the interpolator units defined the order of interpolation being used—from first to eleventh.

It was possible to use other methods of interpolation, such as central differences, but in that case the programmer would have to program all of the individual steps in the calculation, because the built-in controls covered only the Taylor-series approach. The interpolator tape units could also be used, instead of the card feeds, to introduce into the calculator a set of arbitrary constants or parameters.

Use of Iterative Methods

One method commonly used to introduce functions into the machine was iteration. As I remarked earlier, there was no built-in function for square root. In the early days, we used the logarithm and exponential functions for problems involving square roots. This approach, however, was very time consuming, so we developed, as an alternative, an iterative method. This was the Newton-Raphson approach, applied to determine the square root Y of a number X, by utilizing the formula: $Y_1 = (Y_0 + X/Y_0)/2$. The rapidity of convergence depended upon the closeness of the original estimate Y_0 to the desired root. In those problems in which square roots were repeatedly taken, with successive values of X changing by only small increments, an adequate value for Y could be obtained in only a few steps. (If, for example, Y_0 had a relative error of 0.1—that is, 10 percent—then the relative error in Y_3 would be less than 10^{-10}, and in Y_4, less than 10^{-20}.)

Machine Instructions

Instructions were read into the machine one at a time from the program or sequence tape. In one machine cycle an instruction was read from the tape into a set of "sequence relays," and the tape was then advanced by one instruction. The sequence relays were then used to energize the control relays governing machine operation in the next machine cycle. All programming had to be done in machine language, using only explicit addresses.

Three eight-bit fields made up the instruction word. These were denoted A-Out, B-In, and C-Miscellaneous. The instruction to read a numeric word out of a storage register, transfer it into another register, and add it to (or subtract it from) the word already in the second register, was constructed as follows.

A: address of the transmitting register

B: address of the receiving register

C: no code, if addition; subtraction code, if subtraction.

The machine utilized the instructions one at a time, in the sequence in which they had been punched into the tape.

The eight bits in a field were designated by their columnar position as 8,7,6,5,4,3,2,1. Thus the binary code 01001101 would be written by the programmer as 7431—the decimal digits indicating the columnar position of the binary "ones." On the instruction tape, an instruction word was represented by holes for the binary "ones," and no holes for the binary "zeros." The 72 storage registers were assigned addresses 1 through 74, corresponding to the binary numbers 1 through 1001000. The 60 sets of dial switches were assigned addresses from 741 through 83, corresponding to binary numbers 1001001 through 10000100.

A simple transfer and add operation, out of storage address 621 into storage address 31, could be written as shown in table 2. Here the code 7 in the C column is not part of the transfer and add operation, but tells the machine to proceed to the next instruction rather than to stop.

Today's reader may find it odd that the machine was basically designed to execute one operation—add, subtract, multiply, divide—and then to stop unless specifically instructed otherwise. But given the basic timing and organization of the machine, this execute/stop mode did not slow down the machine, and it permitted operator intervention when required. Some of the variations of the transfer and add operation are given in table 3, with the same storage addresses used above. The last instruction would stop the machine only if the operator had set a certain control switch. Reading out of one register into another register did not clear the first register. The programmer could clear a register by inserting the address of that register in both the A and B columns, as shown in the third example in table 3. This instruction told the machine to transfer "minus the contents of the register" back into the register, thus setting it to zero.

Multiplication or division took three instructions, as shown in table 4. The B column code in the first instruction in each set determined whether a multiplication (code 761) or a division (code 76) was to be carried out.

Table 2

A-Out	B-In	C-Misc
621	31	7

Table 3

	A-Out	B-In	C-misc
Transfer, add, and continue	621	31	7
Transfer, subtract, and continue	621	31	732
Reset reg. 621 to zero and cont.	621	621	7
Transfer, add abs. value, and cont.	621	31	72
Transfer, add, and stop	621	31	—
Transfer, add, & conditional stop	621	31	87

Many variations of the multiplication and division coding were possible. Thus, for example, a 32 code in the B column in either of the first two lines of either set would utilize minus the number in address L or M as the operand. Similarly, a 2 code would utilize the absolute value of the number. In division, the quotient was computed to a precision determined partly by plugging, and partly by special control codes in column C.

The built-in functions of logarithm, exponential, and sine required, when used in their simplest form, 3, 4, and 3 instructions, respectively. For example, the logarithm (base 10) of the number in storage address L was obtained as shown in table 5.

The simplest form of interpolation using data from a functional tape, and built-in interpolation controls, required six instructions. The first three instructions positioned the punched tape, and the next three read off the coefficients from the tape and performed the power-series calculation. The argument, which we assume to be in storage register L, had to be read in twice. Using interpolator unit I, the programmer would write the program as shown in table 6.

The number of arguments on the tape, and the number of positions per argument, were set up in special sets of switches. Also, certain plugging was required.

Table 4

	A-Out	B-In	C-Misc.
Multiply no. at address L, by	L	761	—
no. at address M; send product	M	—	—
to address N, & continue.	—	N	7
Divide no. at address L into	L	76	—
no.-at address M; send quotient	M	—	—
to address N & continue.	—	N	7

Table 5

	A-Out	B-In	C-Misc.
Compute log no. at address L,	L	762	—
deliver result to address M, &	831	M	7
reset log input register	—	—	763

To compute a function (such as the square root) using an iterative process, the programmer needed to program explicitly all the arithmetic operations in every step of the iteration.

The reading in of numbers from one of the two card feeds, the printing of results on one of the two typewriters, and the recording of results using the card punch were carried out using special operation codes. Simple examples of the required instructions are shown in table 7.

A number to be printed or punched had first to be read through the bus into the appropriate input register before the printing or punching could proceed serially at its own speed. Card readers, how-

Table 6

	A-Out	B-In	C-Misc.
Read arg. from address L, selecting interp. unit I	L	7654	—
Pick up interpolation sequence control relay	—	—	62
Initiate tape positioning	841	—	—
Read arg. again & perform interpolation calculation	L	763	—
Result to address M, and reset interp. control register	—	M	73
Continue	—	—	7

Table 7

	A-Out	B-In	C-Misc.
Read from L into print reg. I	L	7432	—
Print, and start next. op. before printing is completed	—	752	7
Read from L into punch register	L	753	—
Punch card, and start next op. before punching is completed	—	—	75
Read from card feed I into address M and continue	—	M	7632

ever, which read a word in parallel, were synchronized for direct transfer through the bus to storage registers.

The codes that were used in programming may be briefly summarized as follows. There were 72 address codes for storage registers; these could be used in either the A or B instruction fields. There were 60 address codes for the groups of input dial switches; these were used only in field A. There were also some 20 special address codes that were used, some in field A and some in field B, for the input or output registers associated with the multiply-divide functions, the log, exp, and sin functions, the interpolation procedures, result printing or punching, and various special purposes. There were some 60 operation codes, only a few of which have been discussed here. Operation codes were found in all three fields (A, B, and C); some individual codes could be used in more than one field and could have different meanings in different fields. There were also some program control operations that will be described later. The art of programming the machine evolved rapidly during the early months of the machine operation. Convenience, flexibility, and effective computing speed all increased considerably during this period. Some of these improvements required wiring changes in the machine. Such changes were carried out by Bob Hawkins after we had modified the relevant circuit diagrams that IBM furnished with the machine.

Overall Control of the Computations
The machine was basically designed to carry out a fixed sequence of operations defined by the punched instruction or sequence tape. If the tape was "double-ended," some action would be required by the operator when the machine came to the end. If the tape was formed into a loop (an "endless tape"), the machine would continue to repeat the sequence of operations until stopped by the operator. With suitable programming, the machine could determine when to stop by testing the magnitude of a result variable against a pre-established tolerance. But even when the computation was stopped automatically, the operator still had to take some action such as changing the sequence tape, or moving it ahead or back. As originally built, the machine had only one sequence-tape mechanism. This was a major limitation in the original concept of the machine. During the first several months of machine operation, however, we found that a surprising amount of flexibility could be provided in the form of subroutines and fixed and conditional branching. To facilitate this, we made some changes in the

wiring of the machine, and later we converted one or two of the interpolation tape units so that they could be used for sequence tapes. Even so, a large amount of intervention by the operator was still required. This could involve not only moving the sequence tape ahead or back or changing the tape, but also changing some switch settings, putting decks of cards into the feed or punch, and other operations. Since the machine was relatively slow, stopping it briefly to permit some action by the operator would not necessarily lead to a large reduction in productivity, although it could be a source of error.

In the original concept of the machine, there was no provision for conditional branching, but such capabilities were introduced in the early months of machine operation. One way of providing conditional branching was through the use of the "check counter" (i.e., register). This was a storage register, address code 74, that we specially wired so that it could determine whether the absolute value of a variable was less than another positive number or tolerance. If this condition was not satisfied, the machine stopped so that the operator could take appropriate action. An operation code 64 in field C was used to initiate this control function.

The "choice counter," the storage register with address 732, also had special controls. These enabled one to reverse the algebraic sign of a quantity if and only if some other quantity standing in register 732 was negative. The choice counter could be used to round off a quantity to any desired accuracy, or to handle functions that were discontinuous. An operation code 432 in field C was utilized to enable this function.

Problem Preparation

Before a problem could be programmed for the machine (we called it "coding"), some preliminary steps of preparation were required. These included studying the user's problem in order to gain a clear understanding of the expected results; then selecting the appropriate computational approaches, followed by making the analyses necessary to determine convergence, error control, and other key factors. Some users came in with well-defined problems, and with most of the initial analysis and planning completed; other users came in with only a general idea of what they wanted.

The programmer usually had to break the computation down into a number of separate computing routines, each of which would require a separate program or sequence tape. Tapes that contained programs that would be executed many times would have their ends pasted together to make a loop. Once the program had been written, the instructions had to be perforated into paper tape, and the perforated tape had to be proofread, line by line, against the coding sheets. The programmer had to define all numerical inputs, including computational parameters needed to control the machine. The computations would normally be structured to include suitable checking procedures—for example, obtaining the result by two independent methods, or substituting solutions into the original equations. Often extensive manual, or perhaps automated, calculations had to be performed to establish initial values or input parameters, or to provide sample results to use in checking the program. Decks of punched cards, as well as numeric tapes, had to be prepared as needed.

The programmer next had to provide complete instructions for the machine operator. The instructions would specify all manual operations that had to be performed before a machine run could be initiated. These included setting input constants and control parameters into the appropriate dial switches, wiring up all plugboards that would be used, and mounting card decks in the feeder(s), numeric tapes in the interpolator unit(s), blank cards in the card punch, and paper in the typewriter(s). Finally, the sequence tape to be used initially had to be mounted, and set to the correct initial instruction. The instructions to the operator also had to state the order in which the various sequence tapes would be used, how each tape should be positioned initially, and when during the course of the calculations to move a tape forward or back or change to another tape. The instructions also needed to give the operator some idea (in some cases, a very precise idea) of the results to be expected. In initial machine runs to debug the program, the programmer would usually be on hand to assist in the machine operation and to review early results.

Modifications to the Machine

The characteristics of the machine as described above portray its configuration essentially as it was first put into operation at Harvard in the spring and summer of 1944. A large number of important enhancements, including several major changes, were introduced over a period of several years. Chief among these were the following:

• introducing program branching through the check and choice counters

• providing for calculations at double precision (46 digits) and half precision (11 digits) by using specially configured storage registers and appropriate instructions

• adding a self-standing sub-sequence unit providing 220 lines of instructions determined by a plugboard panel and controlled by relay-operated stepping switches

• adding a high-speed multiplying unit using electronic techniques, and removing the original multiply-divide unit (suitable programming for calculating reciprocals could replace division)

• adding 24 more storage registers to increase the total from 72 to 96

• converting at least one of the three numeric (interpolation) tape units so they could be used for reading sequence tapes, greatly improving the flexibility and overall control of machine processing

• increasing the reliability of the machine by replacing all the original wire contact relays with relays using improved contact materials.

Early Operation of the Machine

Aiken's Interest in Computation

Howard Aiken's interest in the automatic calculator project was by no means limited to machine development, design, and construction. From the beginning he had a major interest in solving problems by computation, as well as in teaching and research. These interests, together with his leadership and entrepreneurial skills, enabled him to establish a multifaceted laboratory, in a very short period of time, once the machine had been completed. He readily admitted that his promotional activities and his close involvement with industry were not in the "classic" Harvard tradition. But with the support of a few key people at Harvard, the major participation of IBM, and, as Aiken discovered (and promoted), the interests of the Navy in better means of computation, he had the ingredients that he needed. And of course the war gave a special urgency to the Navy's computational needs. The confluence of these three organizations to establish the Harvard Computation Laboratory was highlighted during the dedication of Mark I on 7 August 1944. Ironically, there was a falling out between Harvard and IBM (more specifically, between Howard Aiken and IBM's president, Thomas Watson) at the time of the dedication. This led to a break that was never fully mended.

Having sold the Bureau of Ships on the utility of using Mark I in a Navy computing facility, Aiken arranged to be transferred from the Naval Mine Warfare School to serve at Harvard as director of the Navy operation. As I recall, he was transferred in May 1944, though he did pay the Harvard facility some visits earlier in the spring. Before his transfer, the machine had been completed and tested at Endicott, moved and set up at Harvard, and brought into operation on its first two useful problems.

Debugging and Testing in Endicott

The debugging and testing of the machine at Endicott had proceeded in stages during 1942 and early 1943, with each major unit of the machine being tested as its assembly was completed. An early test problem that utilized many of the principal machine functions was run in January 1943. The final work on the calculator was slowed during this period because of the higher priority of other IBM projects, particularly defense programs, but during 1943 the machine was demonstrated many times to groups of IBM personnel and customers.

In the fall of 1943, I selected and programmed five test problems to demonstrate the capabilities of the machine, which was completely assembled and operational at Endicott. Actually, the "problems" were more in the nature of illustrative algorithms or subroutines. They were chiefly intended to exercise all of the built-in arithmetic and functional capabilities. The five problems were given the following titles:

an orbital calculation (with parameters for the planet Neptune).

current in the load of a transmission line

focal length of a plano-convex lens

catenary shape of a suspension bridge cable (with approximate parameters for the Williamsburg Bridge in New York City)

a "math problem" (this was actually the calculation of the log of the gamma function using Stirling's formula)

On 8 December, the machine was demonstrated to a delegation of Harvard faculty members. Plans were made at that time for moving the machine to Harvard. After final checking, it was disassembled for shipment on 31 January 1944. At Harvard it was set up in the basement of the Physics Research Laboratory. IBM sent an installation team of five men to complete the job, and the machine became operational on 15 March. Ben Durfee of IBM, who had done a large part of the detailed design and testing, came to supervise the installation and to assist in the early running of the machine.

The First Real Problems

In the period from mid March through early May, the Harvard team that was responsible for the machine included Robert Hawkins, who had spent a year and a half at Endicott learning how to use and maintain the machine, and who was now in charge of actual operation and maintenance; David Wheatland, a Research Associate at Harvard, who enthusiastically helped us with all aspects of the work; and myself as analyst, programmer, and operator. (I was still involved with the machine only part time, however, as I was still teaching in the Pre-Radar course.) Harry Mimno maintained an overall supervisory interest in the calculator. I remember a discussion with him about what to call the machine—he suggested "supercomputer" in analogy to "superheterodyne."

The first useful applications of the machine were in response to the computational needs of two Harvard scientists. Ronald L. King, Associate Professor of Applied Physics, who was interested in antenna theory, wanted to have some relevant functions evaluated. James G. Baker, an astronomer and a junior fellow in the Harvard Society of Fellows, needed some calculations to aid in the design of a compound lens system. I held meetings with these users, both of whom already had their problems well formulated. We discussed in some detail how the calculations should be carried out. Then I programmed the problems and, with the help of Bob Hawkins and Dave Wheatland, carried out the preparational steps, made initial runs to debug the programs and check results, and completed the necessary computations.

The antenna calculation for Ronald King, which we denoted Problem A, was carried out first. This involved evaluating both a "sine-integral" function and a "cosine-integral function" that were useful in determining the mutual impedance between two parallel coupled wire antennas, for a number of different antenna lengths and spacings. The integrals were

$$S(a, x) = \int_0^x (1/u) \cdot \sin u \; dy$$

and

$$C(a, x) = \int_0^x (1/u) \cdot \cos u \cdot dy,$$

where $u = (a^2 + y^2)^{1/2}$. Results of this calculation were used in a paper by King and a co-worker, published in the *Journal for Applied Physics* for June 1944. The data that we provided at this time were for only a limited number of cases. Additional calculations were made for King later in the year, but ironically these were of no direct use to him because by that time the machine was being operated as a Navy project and the results were "classified." (King had no Navy clearance.) Later, a much more extensive set of calculations of the sine-, cosine-, and exponential-integral functions were made, with results published in 1949 as three volumes of the *Annals of the Harvard Computation Laboratory*.

The lens calculation for James Baker, Problem B, involved tracing skew rays through a seven-element (14 surfaces) telephoto lens system of 40 inches focal length, in order to determine the effect of parameter variations on imaging characteristics. (Tracing skew rays, unlike axial rays, was then a very slow process, because standard rotary or desk calculators were the only digital computing instruments available.) The design was being optimized for manufacture by the Perkin-Elmer Company, to meet the requirements of the Army Air Corps. Baker provided the necessary equations. To take the needed square roots, I decided to use the built-in interpolation process and a punched value tape in order to get a first approximation, followed by Newton-Raphson iteration to obtain the needed precision. There were two kinds of programmed checks utilized.

I still have a few of the papers documenting this work. An early print out of the value tape is dated 4 April, and early problem runs are dated 24 and 25 April. The problem as programmed used four sequence tapes, some 30 storage registers, and five input constants set into dial switches. Sets of three additional parameters were introduced via a card feed. The laboratory did work for Baker over a period of several months, and a number of changes were made in the computing procedures during this time.

The third application of the calculator, Problem C, was a problem for the Bureau of Ships. This was programmed during the late spring of 1944. It involved the correlation of four key mechanical properties of steel (such as tensile strength and Young's modulus) with the concentrations of ten small and uncontrollable impurities in the steel. Once the correlations were determined, the impurity levels in a given lot could be measured, and the steel then allocated (insofar as end-use priorities allowed) to the use most appropriate to its calculated mechanical properties. This problem required solving ten simultaneous linear equations, calculating multiple regression coefficients and multiple correlation coefficients, and making two tests of significance: Snedecor's F and "Student's" t. (The F function is used to test whether the variance between groups is significantly different from the variance within groups. The t function is used to estimate the reliability of a regression coefficient.)

The properties of the reciprocal matrix of the normal equations were such that ordinary elimination of variables could be used in making the inversion. The necessary square roots were calculated by using the built-in logarithm and exponential functions. The original independent and dependent variables were contained in a card deck, with two variables per card, and thus seven cards per sample. Some nine additional card decks were created during the required computer runs. The program was contained on one endless tape and five double-ended tapes. Obviously, setup and operating procedures were much more complicated than for Problems A and B. I did the initial programming and some of the early debugging for the problem and then turned it over to Richard Bloch. He completed the debugging and oversaw the required computer runs.

Conversion to a Navy Operation

By sometime in June 1944, Aiken, now a Commander in the Naval Reserve, had assumed direction of the Harvard Computation Laboratory, which began to be operated as an element of the Bureau of Ships. The laboratory reported to Commander David Ferrier, who headed the Navy office responsible for the Radio Research Laboratory, which was also housed at Harvard. In practice, however, Aiken was able to operate rather autonomously. I began working full time in the laboratory, and in July I joined the Navy as an Ensign in the Naval Reserve. (Aiken had given me a choice between staying as a civilian with deferment for critical war work and joining the Navy. I chose the latter option, though this meant that I was junior to many of the officers who soon joined the staff.) Hawkins, in charge of maintenance, and Wheatland, providing overall support, continued their work. In late April we were joined by Ensign Richard Bloch, USNR, who was transferred from the Naval Research Laboratory, outside Washington. Ruth Knowlton, a civilian, joined us as secretary. In July, Lieutenant Grace Hopper joined the staff, and shortly afterward Lieutenant Commander Hubert Arnold. At about this time, four enlisted men, who all were Specialists (I), First Class (that is, specialists in the use of IBM equipment) joined the group. These men—Charles Bissell, Delo Calvin, Hubert Livingston, and Durward White—quickly took over the operation of the machine, after training by Campbell, Bloch, and Hawkins. Frank Verdonck, Yeoman First Class, also joined the group, to handle Navy administration.

Bloch quickly became the major programmer for the laboratory. He completed the debugging of Problem C, as mentioned earlier, and did the programming for two of the next major problems—K and L. Problem K involved the solution of a set of nonlinear partial differential equations relating to shock wave propagation. Problem L generated a massive (really monumental) set of tables of Bessel functions.

During this period, I prepared with Aiken a set of programming (we called it "coding") and plugging instructions for the machine.

Dedication of the Machine

The next major event was the formal dedication of the machine on 7 August 1944. At that time, IBM formally gave the machine to Harvard, together with a check for $100,000 to aid in its maintenance. The Navy was also a major participant in the dedication, since the laboratory was being operated as an element of the Bureau of Ships. The ultimate importance of this event was the unfortunate rift that it caused between IBM and Harvard. A press release put out by the University News office covered only briefly the relation between the IBM engineers (Lake, Hamilton, and Durfee) and Aiken, and did not emphasize the major role played by existing IBM components and technology in the implementation of Aiken's concept. It also did not emphasize or even formally acknowledge IBM's generosity in designing and building the machine and donating it to Harvard. IBM felt that it had been treated quite unfairly, and angrily withdrew from further collaboration with Harvard, although IBM did continue some technical support for the machine. The feelings between Watson and Aiken were particularly bitter, and were never fully healed. I should say at this point that many of us on the staff of the laboratory also felt that IBM had not been treated fairly.

In the years immediately following the split, Harvard designed and built Mark II, a relay calculator, without using any IBM components. And IBM went ahead on its own to develop and build the SSEC, a dominantly electromechanical machine, with some electronic computing units. Two of the three principal engineers responsible for the SSEC were Frank Hamilton, a developer of Mark I, and Robert R. (Rex) Seeber, who worked with Mark I at Harvard for a period.

Character of the Organization

During the latter part of the summer of 1944, the staff of the laboratory was augmented by four reserve officers: Lieutenant Edmund Berkeley, Lieutenant Harry Goheen, Lieutenant (jg) Brooks Lockhart, and Ensign Ruth Brendel. Rex Seeber, a Navy civilian, also joined the technical staff. Another key Navy addition was William A. Porter (CEM), who was put in charge of the four machine operators. Subsequently, other people were gradually added to the organization, primarily to work on the design and development of the next machine, Mark II.

Grace Hopper was responsible for programming some of the early problems. But before long Aiken made her the principal author of a massive Manual of Operations for Mark I. This documented in great detail the design, operation, and programming of the machine.

The organization was at first rather informal. People helped where they could and did a variety of jobs. Aiken was very much a hands-on manager and participated directly in much of the work. I remember that Mrs. Aiken (Agnes Montgomery Aiken, whom he called Monty) came in to help with punching and checking sequence tapes. It was Aiken's interest, energy, and drive that molded the laboratory into an effective operation, and his imagination and initiative that shaped the laboratory's future courses of action. Below Aiken, except for the operators being administratively supervised by Porter, there was not much structure—Aiken tended to interface with everyone directly.

Although officially addressed as "Commander," Aiken had been known by the early laboratory members (e.g., through Dick Bloch) as "Howard," and we continued to interact informally with him in some respects. Aiken formed strong (either positive or negative) feelings about a new staff member based on his or her initial attitudes and early performance. A staff member who was regarded negatively at first found it difficult to escape from this characterization. One of the naval officers suggested to Aiken that he should read the book *How to Win Friends and Influence People,* but it is doubtful whether Aiken followed this advice. Aiken was very demanding of himself and of everyone else on the staff, but he was very supportive of anyone who he felt was doing a good job. He could also be very cutting to someone who he felt was not.

Very soon after the four Specialists (I) came on board, Aiken began to have the machine operated 24 hours a day, seven days a week, whenever there were problems ready to run. As soon as the Bessel Function Problem (Problem L) was programmed and in successful operation, it served for many years as the base load problem. The first of some 13 volumes of Bessel and Hankel functions was published in 1945, and the last, in about 1951. Extensive tables of other functions— the sine, cosine, and exponential integrals, inverse hyperbolic functions, the error function and its derivatives, etc.—were published between 1949 and 1955. The characteristics of the machine, including its ability to produce camera-ready tables directly, made it ideally suited to this kind of work.

In the early months, operating the machine on a continuous basis was not at all easy. An incoming problem would be assigned to a member of the technical staff (officer or civilian) for overall management and programming. A "duty officer" and an operator would be assigned for each night shift. But the problem manager was also subject to call any time that his or her problem was being run, regardless of who "had the duty." There were programming bugs, errors in setup procedures or input data, and operator errors (sometimes caused by unclear operational instructions) to contend with. But the biggest problems in the early months were errors made by the machine. Often these were intermittent and hence difficult to track down. Sometimes, while one source of error was being traced, a second error would appear. The errors had two principal causes: excessive contact resistance in the relays, and poor contacts in some of the wiring connections.

The relays had "piano wire" transfer contacts, making contact with brass open and closed terminals. At best these provided a relatively high-resistance circuit, and the resistance often became higher with continued use. It was difficult to adjust the relays to provide enough "wipe" of the contacts to prevent contact resistance from building up. (It was my understanding that war priorities made it difficult to obtain more suitable materials during the manufacture of the relays.)

The relays plugged into mounts, to which the interconnecting wires were attached by a crimped connection. Many, but not all, of the connections were crimped only once rather than twice. The singly crimped connections proved unreliable.

Both of these design features were major sources of machine errors, many of which were intermittent and very difficult to find because the error would not always be reproduced. In a few months, all of the singly crimped connections had been replaced by doubly crimped ones. The poorer relays were cleaned and readjusted or weeded out. These changes gradually led to a large improvement in overall reliability. After less than two years, the machine was routinely producing useful results (Aiken called it "makin' numbers") 90–95 percent of a 168-hour week. Eventually, with better materials available after the end of the war, the original relays were replaced with new ones having better contact materials.

Another important element of the early work in the laboratory was the rapid development of new and improved programming and operating techniques. Standard subroutines were established for frequently used procedures. Means for debugging programs and for verifying the correctness of operational setups were improved. New functional capabilities were made available through a combination of hardware changes and introduction of new operational codes.

Postscript

Mark I was used effectively as a computer resource at Harvard for a surprisingly long period. This happened although the machine was soon eclipsed in speed and storage capacity by other machines, and although Mark I was relatively difficult to program and required a relatively large amount of manual intervention by the operator. In 1946 it was moved from the basement of the Physics Research Laboratory to a prominent place in the first-floor machine operating room in the newly completed Computation Laboratory building—later called the Howard Hathaway Aiken Computation Laboratory. Later the Mark IV electronic computer was also installed in the same room, and it gradually took over the main computing workload. Mark I was not completely retired until 1959.

My own participation in the programming and operation of Mark I gradually lessened during the fall of 1944, when I began working with Aiken on the conception and early development of Mark II. During the early part of 1945 I began working on Mark II full-time.

During its period of use Mark I solved a surprising diversity of problems. Oddly, it continued to be used through the 1950s as a teaching instrument. The improvements introduced to Mark I in the late 1940s and the 1950s were major creative design efforts by staff members and students. It is a tribute to Aiken's machine concept and design inputs, to IBM's engineering and construction, and to Aiken's persistence in using his first "computin'" machine that it could be employed effectively over so long a period.

Constructing the IBM ASCC (Harvard Mark I)

Charles Bashe

This chapter supplements the companion volume, Portrait, *by presenting the steps of authorization and construction from an IBM point of view. All the letters, interviews, brochures, and memoranda cited in the text and listed at the end of the chapter are in the IBM archives.*

Aiken and Bryce

James W. Bryce had little reason to believe in November 1937 that a letter from Howard H. Aiken of Harvard University requesting a meeting to discuss "automatic calculating machinery for use in computing physical problems" would lead to anything beyond a recommendation concerning the most suitable IBM machines for Aiken's requirements. Bryce, IBM's leading inventor, was the chief scientific and technical consultant to Thomas J. Watson, the corporation's president. In 1937, the main line of products of the International Business Machines Corporation included tabulators (accounting machines), sorters, collators, reproducers, and multipliers. Each of these machines was designed to perform a fairly specific task in the overall process of maintaining files and preparing administrative reports with the aid of punched cards. Smaller, separate product lines included electric typewriters, electric clocks and attendance recorders. All IBM's products were mechanical and electromechanical; none included vacuum tubes or other "electronic" devices. The corporation had just over 10,000 employees in 1937, the first year in which its gross revenue exceeded $30 million.[1]

Bryce replied promptly to Aiken, and the two met on 10 November 1937 in Bryce's office at IBM's headquarters at 270 Broadway in New

---

1. IBM Yesterday and Today (IBM brochure, 1981).

York. Aiken described problems in the physical sciences that presented vast computational requirements—cases, for example, where the applicable differential equations led to long expansions or to the evaluation of extensive sets of equations. Solving such large problems remained impracticable not because of gaps in theory but because of the time, the cost, and the complex checking procedures involved in producing error-free solutions with batteries of desk calculators such as were then available. After listening to the Harvard man, Bryce recommended that he become fully familiar with standard IBM punched-card machines. Although nearly all installations of such machines were justified by accounting and other "business" applications, a few were in use in the scientific community. Their most notable American scientific user was the astronomer Wallace J. Eckert of Columbia University. In the early 1930s, Eckert had persuaded IBM to donate equipment to the university, including a large and elaborate "calculation control switch" by means of which several machines could be directed through a sequence of quite different operations without the usual requirement of rewiring their "plugboards" (control panels). In 1937, Eckert's computing laboratory was named the Thomas J. Watson Astronomical Computing Bureau.[2]

Planning

On 22 December 1937, about a month after Aiken undertook a course of instruction under the direction of the manager of the IBM Boston office, Bryce arranged a visit to the Endicott laboratory, where Aiken spent four days starting on 31 January 1938. Aiken's host during that visit was Wallace W. McDowell, assistant to the vice-president in charge of engineering, who a few years later would become the first manager of the Endicott laboratory to hold that title. At that time, the company's small group of "inventors"—the senior engineers, in effect—reported only to T. J. Watson, and they defended their status and prerogatives with great care. The importance given by McDowell to Aiken's pro-

2. W. Eckert, interview conducted by L. Saphire, July 1967. See also Jean Ford Brennan, *The IBM Watson Laboratory at Columbia University: A History* (IBM, 1971). Aiken visited Eckert at Bryce's suggestion in order to find out whether the machines installed there would serve his computational needs. For details, see *Portrait*. At a meeting of senior engineers at Endicott in June 1946, Watson recalled: "Dr. Aiken went to Columbia and talked with Dr. Eckert, who is now with us, and then came down to IBM, and we turned the job over to Mr. Bryce and Mr. Lake. . . ."

posal can be gauged by the fact the person to whom McDowell introduced Aiken was Clair Lake, who as one of the company's foremost inventors and senior engineers was entitled to one of several private corner-office areas in the laboratory, an assistant engineer, and a sizable group of designers, draftsmen, and model makers. Lake had been responsible for many important innovations in IBM technology from the time that he joined IBM in 1915 until more than 25 years later. At Bryce's recommendation, Lake was made responsible for determining what kind of IBM equipment would best meet the needs expressed by Aiken.

After listening briefly to Aiken, Lake assigned his assistant engineer, Frank Hamilton, to spend enough time with Aiken to understand his proposal. Aiken's four days in Endicott were spent in discussions with Hamilton, who invited others in from time to time. Aiken expressed the opinion that control information for his undetermined complex of computing machinery should come from a punched tape, but there was little time for discussion of size, format, or operation mechanism. Although Aiken had come prepared with notes on all the various mathematical requirements, the discussion during this visit centered on the basic arithmetic operations of addition, subtraction, multiplication, and division; he postponed to a later date any specifics of logarithms, trigonometric functions, and interpolation of functions. George Daly, an experienced product development engineer assigned at the time to Lake's department, explained at length the operation of the multiplying mechanisms then in use in the IBM Type 601 multiplier.

Reporting to Bryce on Aiken's visit, McDowell brought up the advisability of a more formal working arrangement between IBM and Harvard. He wanted it made clear which party was to do what, and where. He was assuming that IBM would provide only "standard" relays and counters of the type then used in accounting machines. Also, he assumed that IBM would want to protect its sole right to the commercial use of all ideas it would be implementing. However, for multiplying and dividing, Hamilton and Lake chose a method devised by Bryce and his assistant A. Halsey Dickinson.[3] This approach, which was being considered for possible commercial use (the Type 601 was

3. C. Lake, Report on the IBM Automatic Controlled Sequence Calculator, 16 August 1944; F. Hamilton, History with Respect to the Harvard Machine, 21 August 1944. See also IBM Automatic Sequence Controlled Calculator (IBM brochure, 1945).

not designed to divide), had been partially implemented in a machine called the "factorial multiplier," a modified 601 delivered a year or so earlier to the Thomas J. Watson Astronomical Computing Bureau.

Very early in the planning stages of the Harvard machine, Benjamin Durfee became involved in the project as Hamilton's assistant. Durfee had joined IBM in 1917 as a "customer engineer" (field maintenance person); he came to Endicott in 1921, and later he designed the electrical circuits for the factorial multiplier. Durfee was to design most of the electrical aspects of the Harvard machine, including relay logic, arrangements for counter readouts, and routing of information. Quite early, Hamilton and Durfee decided that all adding counters, switches, and card feeds should be connected to a common 24-decimal-digit bus for transfer of values between functional units in the machine. This greatly reduced the volume of interconnecting circuitry, since it made use of the same bus for essentially all information transfer. IBM was not accustomed to scores of accumulators and long factors. Durfee's systemization of the wiring task was an essential feature of the machine's success.

In the Harvard machine, a number was represented by 24 decimal digits, 23 expressing magnitude and the remaining digit representing sign. Aiken had requested this number length in order to achieve the precision needed in the scientific problems for which the machine was intended.[4]

In the original meeting in Endicott, Aiken had suggested using Teletype tape for the sequence control. Lake and Hamilton were later able to point out significant advantages in IBM card stock. Taking the narrow dimension of a card as tape width, it was possible to align 24 (round) holes across the tape—the number required to represent values. Value tapes were to be mounted on the interpolator tape readers (so called because they were used in the interpolation of values from function table as well as in reading other data). It transpired, over a period of a year or more, that the sequence tape also used 24 holes, arranged in three groups of eight.

Although some decisions were being made about the intended design, Harvard's versus IBM's responsibilities remained unclear throughout 1938. In March, Hamilton advised A. Halsey Dickinson, Bryce's assistant, that the cost estimate for counter and relay mechanisms for the Harvard machine was approximately $15,000, exclusive of engineering work and of any special parts that might be required

4. Aiken, letter to Bryce, 3 November 1937.

to mount and drive the units. Hamilton was assuming that IBM would supply only standard parts to the project; he had no basis for an assumption about who would complete the design of the system or build it.[5]

It is not clear from the available records what directions T. J. Watson had given concerning the project in its earliest stages, how large he had expected the effort to be, or just when he had decided that IBM would make a gift to Harvard of the end product. It does appear from such documents as are available that the entire project was understood from the start, by all concerned, to be intended as a contribution to Harvard in the interest of science.

A Commitment

On a Saturday evening in March 1938, IBM corporate secretary John G. Philips dined with Dean Harald M. Westergaard, Aiken, and several other members of the Harvard faculty. They discussed who should do what, and where the proposed machine might be assembled. Aiken felt that the wiring and the assembly would have to be done in a laboratory at Harvard. Philips later informed F. W. Nichol, Vice-President and General Manager of IBM, that he thought IBM would have to cooperate in the cost of panels and other materials. Also, in view of preliminary estimates from Bryce, Philips advised Nichol that the project might cost $50,000 to $75,000, of which about half would be for parts.

Bryce attempted to get from Lake a more accurate estimate of the project costs, inquiring whether $1000 would suffice for a preliminary layout to facilitate estimation. McDowell replied for Lake that no accurate estimate could be developed unless there was a wiring diagram. He reminded Bryce of IBM's understanding with Aiken that "all the special work would be done at Harvard" and "all we would supply would be standard units such as counters, relays, switches, etc."—the items covered by the $15,000 figure given earlier. Bryce, in turn, doubted the wisdom of expecting Harvard to cope with details; he felt such steps as mechanical mounting of counters in such a way as to ensure correct operation would require careful checking by IBM.[6]

5. F. Hamilton, telegram to H. Dickinson, 3 March 1938.

6. W. McDowell, memorandum to J. Bryce, 23 March 1938; Bryce, memorandum to McDowell, 24 March 1938.

Late in April 1938, Bryce met in Endicott with Lake, Hamilton, and McDowell. Bryce dickered with McDowell over the amount of money required for a preliminary layout of the Harvard machine. He agreed with a suggestion by Philips that the project should start with $1000 rather than the $2500 proposed by Lake and Hamilton, and he proposed that "by the time Mr. Watson returns to New York" there should be something more definite to discuss. Implicit was the indication that someone at headquarters had begun to assume that IBM would provide, at least, "a sort of diagrammatic layout."[7]

By May 1938, Bryce had sent Hamilton a list of the devices which he and Aiken considered practical and satisfactory to perform the functions of the machine. The list included numbers of switches, counters, and other devices (many of which would change in the course of system design). Among the specifics were the following:

1) On one panel, a keyboard for entering independent variables into four counters of twenty-four digits each, along with permanent "setups" for twenty-seven twenty-four-digit values.

2) On another panel, fifteen (twenty-four-place) "parking counters" and four sets of adding and subtracting counters.

3) Twenty-four sets of (twenty-four column) hand setup devices (again on a separate panel).

4) Two sets of multiplying devices (factors and results both being twenty-four places).

5) A "difference engine" with four sets of counters for *each* of five orders of difference; the capacity within each set decreasing from twenty-four columns for the first order to five columns for the fifth order.

Relays, special switches, plug hubs, and power panels were mentioned as occupying other panels. All registers were to be capable of serving as sources or destinations for data to or from other registers or external devices such as punches or printers. "This is going to call for a large number of plug connections," Bryce noted, "and we will have to provide some master shaft for timed contacts."[8] Bryce's description was not explicit that the "multiplying devices" would be used for direct division, but other records indicate that Lake and Hamilton made that

7. Bryce, memorandum to McDowell, 5 May 1938.

8. Bryce, letter to Hamilton, 16 May 1938. There appears to be no indication as to who suggested the use of Charles Babbage's term "difference engine." Bryce owned a copy of Babbage's autobiography, which he later presented as a gift to Aiken (for details, see *Portrait*).

determination very early, if not during their first four days with Aiken at Endicott.[9]

In May 1938, Lake was authorized by work order to prepare the panel layouts. Work began in August. A preliminary layout was completed in September, and Bryce, Lake, and Hamilton spent three days in Lake's office arriving at a cost estimate of $100,000 for the machine. During the second half of 1938, the organization of the machine was still unclear, especially with regard to whether the various functional units would be coupled mechanically or synchronized electrically. This was discussed at some length by Aiken and Bryce, as was the question of how best to specify an authorization for actual design and construction. Bryce took the question of authorization to C. R. Ogsbury, Watson's executive assistant, and in January 1939 he provided Ogsbury with a two-page description of the project[10] that included the following paragraph:

With regard to construction, it is proposed to use standard accounting machine parts so far as possible. These parts are adding counters, relays, plug wires, switches, etc. Because the machine is not really one machine, but a series of computing machines, flexibly coupled together in a large number of combinations, it is proposed to erect the various parts in the building where they are to be used. The computing and controlling mechanisms will be assembled on panels and provided with suitable mechanical driving and electrical coupling circuits. The panel mountings, mechanical drives, etc. will have to be made specially for this job. Most of this work will have to be done by IBM. Some of the special mechanisms for record control of the machine can undoubtedly be done at Harvard from designs made by IBM. An accurate estimate of the total cost is impossible at this time, but an estimate of the standard parts necessary places this at approximately Fifteen Thousand ($15,000) Dollars. On account of the amount of planning and special work necessary and based on experience it is estimated to complete the machine, the total cost will probably be from Seventy-five Thousand ($75,000) Dollars to One Hundred Thousand ($100,000) Dollars and would be spent over a period of approximately two years time.

The wording of Bryce's descriptive note reveals an IBM commitment taking shape somewhat more rapidly than the visualization of the final machine. Even for Bryce, it was still more natural to think of the system as "a series of computing machines" rather than as an automatic calculator or computer. "Systems" of IBM equipment, up to that time, had consisted of individual punched-card machines, each

9. Hamilton, History with Respect to the Harvard Machine.
10. Bryce, letter to C. R. Ogsbury, 18 January 1939, with attachment.

designed to perform one or few operations upon all cards of a deck fed through in one continuous pass. With a few exceptions (the most important being Wallace Eckert's system at Columbia), the flow of operands and intermediate results was accomplished by operators who carried decks of cards from one specialized card device to another. System control resided in written or memorized procedures followed by skilled operating personnel and in the plug wiring of a machine's removable control panels. But the system envisioned by the beginning of 1939 to meet Aiken's requirements provided for a multiplicity of basic arithmetic operations and selected mathematical functions, all orchestrated by a common sequence control. Whether regarded as a machine or as an assembly of machines, it distinctly foreshadowed the computers of the electronic era.

Early in February 1939, Bryce informed Dean Westergaard that, since Mr. Watson had just authorized IBM to proceed with the project, Bryce would have a contract drawn up to clarify responsibilities. By the end of March, the Executive Committee of IBM's Board of Directors had approved an agreement; by May 1939, the document and attachments that provided patent rights to IBM had been signed.[11]

The Project Is Underway

In May 1939, Lake received an appropriation of $15,000 and an authorization to proceed with design and construction of the Harvard machine. Lake and Hamilton soon spent a day with Aiken and Bryce in the latter's office in New York and managed to carry the proposed specifications forward somewhat. Aiken then spent much of the summer of 1939 (from 27 June through the end of August) at the Endicott laboratory working with Lake, Hamilton, and Durfee and learning from them the many uses of relay contact points and emitters.

An improved estimate, completed late in September 1939, placed the cost at $85,000. Lake had been warned not to exceed a project cost of $100,000 without specific authorization from Mr. Watson. Hamilton, feeling that the machine should be trimmed to permit a contingency of 80–100 percent (that is, to a cost of $50,000 or $60,000), deemed $85,000 too close for comfort.[12] He discussed this with Lake,

11. The Harvard-IBM contract, with attachments assigning patent rights to IBM, is in the IBM Archives. Another copy is in the president's papers in the Harvard University Archives.

12. Hamilton, History with Respect to the Harvard Machine.

Bryce, and Aiken in Bryce's office in October 1939, and they agreed to removal of the entire "difference engine," several panels of multiplying-dividing equipment (a single M-D unit remained to serve all needs[13]), one of the tape-reading mechanisms, and the cost to IBM of a power panel and a mechanical drive motor (which Aiken agreed that Harvard would furnish).[14] In November, Hamilton wrote Aiken that his engineers had nearly completed the revised multiply-divide diagram and that all functional diagrams had been completed and checked. From the start, Lake had looked to Hamilton for day-to-day management of the project; as final design and construction got underway, Hamilton was effectively in charge.

Aiken and Endicott

In 1940, Aiken again spent his summer at the Endicott laboratory. Final details of the design were being completed, and construction was already underway. Aiken spent most of his time checking on changes that had been agreed to the preceding October; however, according to Hamilton, he also taught a math course ("without remuneration").[15]

There is surprisingly little about Howard Aiken in IBM's records or in interviews with IBMers who interacted with him. This paucity of information may be due to the project's unpopularity in the laboratory. Engineers were not striving to be assigned to it. No one was going to gain recognition in the traditional way, since no great revenue (in fact, none at all) would result. The project was viewed as compliance with Mr. Watson's wish to make a contribution to the scientific community as represented by Harvard. In addition, there had always been an atmosphere of secrecy in the IBM laboratory, largely because Watson encouraged inventors to compete with one another. It is unlikely, therefore, that anyone in the laboratory outside Lake's domain had much information about the project. Most people were aware of its existence, of course. George Daly remembers that at its peak the project occupied several dozen people, including draftsmen and modelmakers.[16]

13. B. Durfee, oral-history interview conducted by L. Saphire, November 1968.
14. Hamilton, History with Respect to the Harvard Machine.
15. Ibid.
16. G. Daly, interview with C. Bashe, December 1981.

In March 1941, Aiken sent Hamilton the values of 36 logarithms and 21 sines to be stored in the machine, also outlining in detail the requirements for interpolation. In mid April, Hamilton notified Aiken of an idea he and Durfee had had for incorporating a counter to determine the final tape position, suggesting a means for keeping track of that position and asking for advice about the capacity of the counter. These and other design details were discussed in May, when Aiken made his last visit to the Endicott laboratory before being called to active military duty. Apparently having satisfied himself that things were proceeding according to plan, Aiken notified Lake and Hamilton that he would have very little more time to spend on the machine, and he designated Robert Campbell of Harvard as his representative. (For details, see *Portrait*.)

True to his word, Aiken became uninvolved and difficult to reach. In February 1942, Hamilton and Campbell visited him at Yorktown, Virginia, and discussed progress. In March, after Campbell transmitted to Hamilton the values of the powers of 10 needed for computing antilogarithms, the values were promptly wired to the machine. In 1942, Hamilton's group worked primarily on the construction of the Harvard machine. By then, however, the project was subject to disruptions caused by higher-priority projects related to the war. Around September of 1942, Hamilton was assigned to build a set of special instruments for the Navy, and much of his time was spent on that project, though he remained interested in the Harvard machine and spared what time he could for it. During 1943, because of Hamilton's wartime duties, the burden of testing the machine fell heavily on Lake and Durfee.

Completion and Dedication

The Automatic Sequence Controlled Calculator solved its first practical problem on 1 January 1943. The New Year's holiday date hints at the demands that fall upon a development group when a large investment of money and effort over a period of several years has finally been brought to fruition. The resulting complex of machinery cannot be left idle; efficiency demands that the schedules of individuals be adapted to maximizing the use of the newly assembled machine.

The first problem given to the machine involved finding the current in an inductive electrical circuit as a function of time. It required using the capabilities of the machine to add, subtract, multiply, divide, and compute logarithms and antilogarithms. Hamilton sent a description of the problem, and its solution in graphical form, to Lieutenant Commander Aiken, but he never received an acknowledgment that Aiken had received it. The entire year 1943 was occupied with testing the machine and making corrections as required. During the latter half of 1943, the machine was demonstrated many times to IBM's Customer Administrative, Customer Engineering, and Systems Service Schools. On 7 December, a demonstration was given to visiting members of the Harvard faculty, including President James Bryant Conant, Professors Harlow Shapley and E. L. Chaffee, and Howard Aiken.

The giant machine was disassembled, and on 31 January 1944 it was brought to Harvard on several trucks. The next day, when it arrived, a number of Lake's men were on hand to unpack and assemble it. The machine was placed in operation on 15 March.[17] In mid April 1944, Conant informed Watson that the machine was "already being used for special problems in connection with the war effort" and that he would let Watson know about the plans for the dedication. That gathering was later set for 7 August, and Watson was invited to "say a few words." Watson set out for Harvard on 6 August, looking forward to making a monumental gift to science and education in the name of the IBM Corporation.

The event was an acute disappointment to Watson. In Harvard's news release and in a press conference on the eve of dedication, Aiken billed himself as *the* inventor whose original conception had been transformed into a completed machine after six years of design, construction, and testing. The assistance of Lake, Hamilton, and Durfee and the role of IBM in general were mentioned but given little prominence in the press release. After sharp words with Aiken, Watson participated in the ceremony as had been planned and presented the ASCC as a gift from IBM to the university, along with a check for $100,000 for operation of the computing laboratory. The machine continued in operation at Harvard until 1959, but there were no further joint projects between the university and IBM.[18]

17. T. J. Watson, letter to J. B. Conant, 20 December 1943.

18. I am indebted to Lyle R. Johnson for many valuable suggestions during the preparation of this chapter.

Programming Mark I
Richard Bloch

This presentation is based on the author's personal recollections, notes, and programs, supplemented by volume I of the Annals of the Computation Laboratory of Harvard University, A Manual of Operation for the Automatic Sequence Controlled Calculator *(Harvard University Press, 1946; published by The MIT Press in 1985 as volume 8 in the Charles Babbage Institute Reprint Series for the History of Computing). See also Richard M. Bloch, "Mark I Calculator," in* Proceedings of a Symposium on Large-Scale Digital Calculating Machinery *(Harvard University Press, 1948; MIT Press, 1985).*

It was December 1943 and I was an ensign in the US Naval Reserve, stationed at the Naval Research Laboratory in the Anacostia area of Washington, D.C. I had been inducted into the Naval Reserve shortly after graduating from Harvard the previous June with an honors degree in mathematics. One afternoon I received a call from the administration office of the Naval Research Laboratory requesting that I escort a visiting officer through various NRL facilities. The visitor was Lieutenant Commander Howard Aiken. In the course of this tour, Aiken inquired as to my educational background and my aspirations for the future. When he learned that I had attended Harvard and that mathematics was my field, his eyes seemed to sparkle. "How would you like it," he asked, "if I were to arrange for your transfer back to Cambridge to work with me on an exciting new project at Harvard under Naval auspices that I am certain will intrigue you?" I naturally wanted to hear more. Aiken told me he had conceived the idea of an "automatically sequenced digital calculator" that would solve all sorts of complicated mathematical problems at high speed. The machine, he indicated, was in the process of being delivered from IBM's plant in Endicott, New York, where it had been built. It was to be set up in the Cruft Laboratory at Harvard, and it would be operational within a few months. This seemed too exciting to warrant any extensive cogitation on my part, and I immediately accepted the proposition.

The New World

In the spring of 1944 I arrived at Harvard's Computer Laboratory. What I saw on my first day there was much more spectacular than I had expected. The machine, about 50 feet in length and 8 feet in height, was an imposing sight. I had some experience with electro-mechanical desk calculators, such as the key-operated Marchant and Friden machines; I also had some acquaintance with various IBM punched-card machines. This Automatic Sequence Controlled Calculator, however, was clearly not of the same species. Its size, its speed of computation, and its capability of proceeding through an enormous series of arithmetic operations automatically without human intervention placed it in a class of its own. What I beheld on that day in March was the ancestor of all the automatically programmed general-purpose digital computers that were to follow.

The Learning Period

The machine had barely been assembled, and the packing boxes that had housed the various units during their trip from Endicott to Harvard were still to be seen in an outer room. Ben Durfee of IBM was still tracing circuits, as was Bob Hawkins, the Harvard electrical and mechanical maintenance engineer in charge. The first programs, generally of a test nature, were being run by Bob Campbell, the man actually responsible for getting the machine into productive operation. Howard Aiken was there when I arrived, and he indicated that he expected me to assist in getting the machine operational. In particular, I was to become fully capable of programming[1] the calculator as soon as possible.

I began by poring over the huge books of blueprints in which all the circuits of the machine were illustrated. Though I had no engineering background, I was aware that it was going to be absolutely necessary for me to understand the workings of the calculator in every detail if I was going to do an intelligent job of programming. Furthermore, I applied the same discipline here that I had observed in my

---

1. In this era there was no usage of the word "program," either as a verb or as a noun, to describe this activity of evolving the set of operational instructions which the machine was to execute. It was to be, as a matter of fact, a few years before the "program" would come into common usage.

studies in mathematics and physics: never memorize a formula or a relationship without understanding the process by which it was derived, and be able to reproduce the derivation from the basic elements. That, I felt, was the only way one could become really expert in any field of endeavor. Fascinated by the logic of the machine's design, I immersed myself in the blueprints for days on end; when I couldn't understand something, I sought help from Durfee or Hawkins or Campbell or Aiken—all were wonderfully patient in resolving my difficulties. Within a month or so I began to feel confident that I understood the inner workings of the design.

As I was to realize shortly, there were peculiarities in the machine's execution of the instructional codes for many of the basic operations. In fact, the programming disciplines that were eventually instituted to optimize performance were intricately related to the logical design of the machine. My persistence in understanding the circuitry and the design logic was soon to pay some important dividends in avoiding programming errors related to internal timing considerations, in tracking down machine failures of various kinds and getting the machine back into productive operation quickly, in resolving the possible confluence of programming errors *and* machine malfunction, and (later) in making design changes and additions to the machine that I was convinced would improve its performance considerably.

Mark I, as Aiken soon named the machine, consisted of a massive quantity of electromechanical devices. There were approximately 3500 electromechanical relays, each having between four and twelve poles. There were some 2300 electromechanical storage counters, hundreds of ten-position mechanical switches, and thousands upon thousands of backwired relay terminals, not to mention the card equipment and the mechanical sequencing and interpolation mechanisms. The frequency of failure of these devices obviously had to be kept extremely low if the machine was to remain operational for any sustained period of time. Fortunately, after some faulty and marginal devices were weeded out during the early days of operation, runs of several hours could proceed unimpeded by a single machine failure. When a failure did occur, it was essential to track down the source with dispatch. It was at this point that I was to discover the value of a thorough understanding of the machine's circuitry as well as an intimate knowledge of the numerical and logical flow of information in the computation being run.

Programming—A New Art

Other than the instructional codes for multiplication, division, and other operations, which were tabulated crudely in a book, there was no manual. There was no precedent for the machine, and clearly there was no precedent for the technique of programming such a machine. Aside from helpful encouragement by Aiken and Campbell, I was on my own in creating tools and techniques for this new art. It seemed that, in Aiken's plan for the laboratory, I would probably be responsible for the programming of a major share of the problems brought to the laboratory for solution by Mark I. There were only three on the professional staff: Aiken, Campbell, and I. Campbell was already beginning to spend much of his time conferring with Aiken on possible parameters for a second machine. Aiken had many other responsibilities, some concerning the establishment of the facility, its Navy sponsorship, relationships with Harvard, and the acquisition of important computational problems pertinent to the defense effort. Thus, I was left to develop and apply the new art of programming and to put Mark I to work solving problems.

Mark I was a fixed-cycle, 23-decimal-place machine. The cycle time was set by the revolution time of the mechanical drive apparatus: 200 rpm. Consequently, the basic cycle was fixed at 300 milliseconds; this was, then, the time it took to read a single line of code. There were 72 storage registers, each capable of storing a 23-digit decimal number and its algebraic sign. The machine was of fixed-decimal-point design: for any given run, the decimal point had to be established between any two adjacent decimal columns. Appropriate plug wiring enabled this decimal point placement to be observed properly for all operations in the machine, including the shifts required in multiplication. Practically every operative unit of the machine required a plug-wire array, which would generally change from problem to problem and even between different machine runs within the context of a given problem. "Plugging instructions," therefore, had to be specified by the programmer for every unit and for every run. Any error in these instructions could foul up a run immediately. Inadvertent "crossing" of wires would have disastrous results, and during debugging a plugging error would sometimes masquerade as a failure of a relay or a storage counter.

Negative numbers were stored as nines complements of the positive absolute value of the number, a nine in the 24th column denoting a negative quantity. Thus, if the decimal point was between columns 20 and 21 (columns were numbered from right to left, the least significant digits being the low-order columns) and we wished to store the quantity –3.4569, that quantity would actually reside in the machine's storage register as follows, the numbers in the upper row representing columns:

24 23 22 21 20 19 18 17 16 15 . . . 1

---

 9 9 9 6 5 4 3 0 9 9 . . . 9

In order that addition and subtraction operate properly, each of the 72 storage registers was equipped with an end-around carry circuit, enabling a carry from column 24 back into column 1 whenever column 24 had a 9 stored and a carry from column 23 occurred. This interesting aspect of the design led to certain additional problems for the programmer. Whenever the plugging called for a shift right (multiplication by 10^{-x}) of a quantity, care had to be taken to supply nines to the left of the most significant digit in the shifted result if the quantity were negative; otherwise some huge negative number would appear, wreaking havoc with the computation. A left shift of a negative quantity did not require the same precaution, since at worst the effect would be a variance of 1 in the least significant digit of the result whether right-hand nines were supplied or not. A curious aspect of this nines-complement representation of negative numbers was the intriguing existence of two different representations of the quantity zero. All nines residing in a storage register would read out as –0.0000 0000 . . . , whereas all zeros would read out as 0.0000 0000. . . .

Each storage register was also an adder. By the very nature of the design of the IBM storage counters (each of the 72 registers contained 24 of these counters), addition was in effect an inherent operation. Each of these counters had ten steady states, representing the ten decimal digits; no power was required to remain in any given state, since the ten digits were actually represented by ten different physical brush-contact points on a commutator ring. A brush resting (say) on the "4" position in a given column would remain in that position until an impulse was given that caused the brush to move mechanically seven positions around the ring if "7" were the digit in the corresponding column of a second quantity being added to this register; as a result, the counter wheel came to rest at the "1" position and the add without carry was effected. Carry circuits would then complete the addition in a single step through all 24 counters, utilizing special nines- and tens-carry contacts on each of the counters.

An addition, a subtraction, or a register transfer occupied one machine cycle of 300 milliseconds. This was also the time required for the main sequence mechanism to read a single line of coding—said line being sufficient to completely define one of these basic operations; therefore, if there were nothing but a sequence of additions or sub- tractions, the sequence mechanism would continue in motion, its read- ing pins sensing the punched holes in the main control tape for successive lines of coding every 300 milliseconds. This, then, was the maximum speed at which a main control (or instructional) tape could be read, the machine executing each line of code in synchronism with the sequence mechanism. The action of the mechanism's steel clutch engaging and releasing at a frequency of some 3+ times per second emitted a distinct sound, not unlike the clatter of steel-shod horse's hooves clanging along a paved street. Since by definition this was the fastest pace at which an instructional tape could move, this sound became the most heartwarming and satisfying auditory sensation that could fall upon the ears of the "programmer du jour"—let alone the entire complement of the Computation Laboratory.

It was practically impossible, in a legitimate computational run, to keep the sequence mechanism in perpetual motion. Many arithmetic operations would require cycles to pass with no sequence mechanism motion; however, even in this event the trained ear would know that the machine was still "crunching" numbers; storage counters by the dozen would be zipping around and banks of relays would be clatter- ing while executing these multi-cycle instructions. When the instruc- tion was complete, the sequencer would once again clang. Howard Aiken's desk was in an office separated from the machine room by a wall consisting of large plate-glass windows. My desk and worktable were in the machine room, directly facing Howard through the win- dows. When Mark I came to a halt in its computational chores for whatever reason, an eerie silence engulfed the machine room. Since generally it was my program being run, I whirred around in my seat with a perturbed look; at the same time Howard, no matter how deeply engrossed in his desk work, would look up and see me heading for the machine's control area and in no time he was out of his office at my side, asking "What's wrong, Dick?" Howard was at peace with the world only when Mark I was producing numbers, and I found the same feeling permeating my being. I often thought that, because Howard firmly believed he had a kindred spirit in me, he treated me in a very special way.

Machine Cycles

When Mark I was designed, it was deemed advantageous to incorporate multiplication, division, and three basic transcendental functions ($\log_{10}x$, 10^x, and $\sin x$) as auto-programmed or "built-in" multi-cycle functions or operations. In nearly all numerical computations, of these five operators, multiplication occurs with the greatest frequency by far; fortunately, in Mark I this operation was also the least time-consuming of this multi-cycle group. The transcendentals required approximately 200 cycles—a minute—to produce the function accurate to twenty-plus places of decimals; as a result, these instructions were assiduously avoided in nearly all applied programming activity. It was indeed ironic that these functions occupied a sizable proportion of the machine's physical complement of relays and associated equipment, and that a great deal of creative effort during the period of logical design surely had been expended in this area.

   Multiplication was coded in the three-address system of the machine as follows:

OUT	IN	MISC
21	761	—
3	—	—
	4	7

The first line, when read by the sequence mechanism, instructed the machine to transmit the contents of register 21 into the multiplication circuitry as the multiplicand. The second line introduced the multiplier from register 3. Finally, the product was to be sent to register 4. The "7" in the MISC column was a special code instructing the sequence mechanism (SM) to read the fourth line (not shown above) of code, act upon it, and step ahead to the fifth line of code. In general, a "7" in the nth line of code instructed the SM to read the $(n+1)$st line and proceed to rest on the $(n+2)$nd line. The design of the machine, however, required that many operations—multiplication included—read and execute the lines of code on specific machine cycles as determined by special sequence circuits controlling that particular operation. Thus the "7" in the third column of certain lines of code was to be omitted by the programmer; and instead a "pseudo-7"—known to the habitués of the laboratory as an "automatic"—would be internally generated only when the proper cycle had arrived for reading the next line of code. In the multiplication example above, once the multiplicand (MC) was read in, the multiply "engine" had to build up the nine multiples of the MC prior to the multiplier being introduced; this required two extra machine cycles to complete; only then would the automatic be generated, causing the SM to read the second line of code and come to rest on the third line. The second line, upon reading in the multiplier (MP), caused the multiplication process to proceed in earnest. Each decimal digit of the multiplier was sensed in sequence, and the corresponding multiple, with appropriate shifting, was sent to the product register—a 46 digit accumulator—until all MP digits had been exhausted, at which time the product was ready for transmission. An automatic was then issued causing the third line of the code to be read and the SM to step to line four; in accordance with the third line of code, the product was sent to register 4, and the multiplication process was complete. Since there was a MISC "7" code in this third line, the SM proceeded to read the fourth line of code—actually the first line of some new instruction to the machine—and step to the fifth line.

Parallel Processing

The total number of cycles required in the multiplication was a function of the number and position of "0" digits in the multiplier. As many as 20 cycles could be consumed in the total operation, and as few as eight cycles if the MP was zero. However, only three of the cycles consumed in the operation required any activity on the main buss of the machine: the cycles associated with the readin of the MC, the MP, and the readout of the product. It was soon realized that, with some care, it was possible for the programmer to interpose a number of operations *within* the multiplication instruction, so long as these operations did not affect the multiplication in progress. In particular, two lines of code could be inserted in the interval between the introduction of the MC and the MP; furthermore, several lines of code could be inserted between the readin of the MP and the readout of the product.

Since my programs often were operating with as many as twenty or more decimal places, my multiplications were of long duration. Accordingly, I would design the program in such a way that a maximum number of operations were interposed. Each line of code interposed represented a cycle of machine time saved during each revolution of the program tape being run; it turned out that a significant speedup resulted from the repeated use of interposition throughout the program.

The programming technique involved forcing the main sequence mechanism to continue to step ahead to a new line of coding by inserting a "7" code in the MISC column of successive lines of code, but being careful to omit the "7" in the line just prior to the MP readin and product readout lines of the multiplication. The result, when properly programmed, was that the SM would always be resting on the proper code line of the multiply instruction at least one cycle prior to the "automatic" being issued internally; thus, all the interposition was completely transparent so far as the multiply operation was concerned. The resultant coding would appear as shown in the following multiplication example (an actual excerpt from one of my main program tapes, one for the calculation of Bessel functions).

LINE NO.	OUT	IN	MISC
282	6	761	7
283	632	87	7
284	641	8731	32
285	82	—	7
286	64	873	732
287	641	871	732
288	64	87	732
289	6421	8731	7
290	642	873	7
291	6421	871	32
292	—	651	7

Lines 282, 285, and 292 constitute the multiplication instruction; all eight of the other lines of code are interposed operations which otherwise would be programmed *external* to the multiply coding. The cycle consumption of the multiply in this instance is 14 cycles with eight cycles interposed. Under standard coding rules without overlap, a total of 22 machine cycles would have been required; consequently, the saving in machine running time amounts to about 36 percent—a significant gain of speed.[2]

Another important source of time compression involved the overlapping of printing—normally a 23-cycle operation—with all other program execution. So long as successive printing instructions were spaced sufficiently apart in the code sequence, it was possible to reduce the net additional time consumed by a print to two cycles. Similarly, it was possible to reduce the card-punching instruction effectively from a ten-cycle to a two-cycle operation, insofar as total elapsed machine time was concerned, by proper placement of the punch instruction within the program.

In the cases of both printing and punching, the overlapping simply enabled either activity to proceed in parallel with other programmed operations.

Several other programming maneuvers that compressed the number of programmed lines of code were instituted. In many instances it was possible to fill an otherwise blank MISC position with a productive one-address command, such as the instruction to step a value tape forward or back. In other cases, what would normally have been blank IN or OUT positions in certain code lines could be filled with a transfer-producing single storage register address, or a one-address command instituting some required action; this stratagem conserved a machine cycle for each such insertion. Some of the new code-compression techniques became quite complicated, especially when instructions with internal automatics were enfolded in the multiplication instruction sequence (which itself, as discussed earlier, generated its own series of automatics).

Overall, I calculated that the best programming compression techniques probably reduced running time for most programs by between 25 and 35 percent. In the case of certain runs this amounted to a saving of days of run time; indeed, weeks were unquestionably saved

---

2. It was also true that the division instruction was subject to interposition using techniques similar to that described above; and because this was an extensive operation in cycles consumed, a great number of interposed lines was possible. However, the paucity of division operations in most programs lessened the overall time saved from this source.

in the extensive tabulation of the Bessel functions. Unfortunately, this compaction of machine code did not come without a price. Although I tried to annotate my coding sheets thoroughly, it was at times almost impossible for an operator running a program to decipher exactly what was going on. Aside from the fact that the logical flow of the program was at times terribly complex to follow, the compaction of code made the task of analyzing and tracking down the cause of a sudden machine stoppage doubly difficult. However, I worked closely with the four or five "I-Specialists"[3] who were assigned to the laboratory as machine operators, and after a surprisingly short learning period they were able not only to run the programs smoothly in accordance with complicated operating instructions but also to be of considerable assistance in debugging in times of trouble.

Phases of Programming

Here I will trace the successive phases of activity required of the programmer from the time a mathematical problem was first brought to the Computation Laboratory through to the final objective—the successful running of the problem on Mark I.

In the first phase, a determination had to be made as to the mathematical attack on the problem. At times the originator of the problem had this matter well in hand; at other times, the end product or the desired results were defined but the identification of the equations to be solved numerically had yet to be specified.

In the second phase, with the equations determined, the precise numerical methods to be applied in yielding the desired solution or functional tabulation had to be chosen.

In these first and second phases, Howard Aiken often would join in reviewing the problem under consideration and participate enthusiastically in analyzing the comparative advantages and disadvantages of various potential methods of solution; certainly Howard's extensive knowledge of numerical methods proved extremely valuable in accelerating my ability to learn and apply this branch of mathematics to the assignments at hand.

3. These men were enlisted Navy personnel who had working experience with IBM card equipment and were accordingly given the special ranking of I-Specialist, with ratings of first, second, and third class based on merit and tenure.

The third phase embodied the overall planning of the program. Among the important issues to be resolved at this time were the following:

inherent numeric accuracy desired in the solution

positioning of the decimal place within the machine

determination of input parameters

determination of output parameters

input media to be utilized (e.g., punched cards, value tapes via interpolators, manual switches)

output media to be utilized (e.g., punched cards, typewriters)

number and organization of main control or sequence tapes

normalization requirements to maintain precision

special shifting requirements, and which special registers to accomplish same

method of evaluating transcendentals, if required (choices included built-in machine instructions, use of interpolators for automatic Taylor-series evaluation, tabular difference schemes, special series expansions).

Certain of the issues listed above involved consideration of human factors, such as ease of operation and possibilities of human error. For example, using punched cards for input to a run obviously required that the cards be punched in a previous machine run or manually; especially in the latter case, the cards had to be checked for error—a time-consuming manual operation. Furthermore, a deck of punched cards was obviously in an ordered sequence, and any inadvertent interchange of card positions could invalidate the entire machine run. If, on the other hand, input quantities were to be punched into a value tape and read by an interpolator (a device similar mechanically to the main sequence mechanism), then, once the tape had been verified as correctly punched (usually via a printout), the sequence of successive values as called forth by the program could never be disturbed; any inversion from one run to a successive run was impossible. Furthermore, if the run was to be repeated using the same value set, the tape was simply revolved *under program control* back to its initial position; with cards, on the other hand, manual refeeding was necessary. One other important point is that the sequence of values read from the tape did not necessarily correspond to physically adjacent values as punched into the tape; it was possible to step the reader mechanism back or forward (unlike the main instruction tape), skipping over values via programmed instructions; with punched-card input, this was impossible. If it was necessary, however, to arrange the output of one run to become the turnaround input to a successive run (and in certain problems this was unavoidable), punched-card input and output was the only solution; unfortunately, Mark I had no punched tape output; also on the debit side of the ledger, punching a value tape was a more laborious and error-prone process than punching cards.

The number and arrangement of main control tapes referred to above was an issue of some importance. A program tape could be made endless by simply overlapping duplicated codes at the beginning and end of the tape, then gluing the ends to form an endless loop; this was done for any program that was to be repeatedly executed through a number of revolutions. Here the identical instructional sequence was executed in successive revolutions, with an incremental change in one or more arguments or input variables automatically inserted at the beginning of each revolution. Often it was necessary to arrange for a separate "starting" tape to be programmed and run through a single pass; this would enable certain starting or initial values to be read in or perhaps calculated and inserted in appropriate storage registers before the main multi-cycle tape could be run. In other instances a problem (e.g., one solving a set of n simultaneous linear equations in n unknowns) was solved in one revolution; in this case, the program tape could contain literally thousands of lines of code to be executed by the machine but once. No matter how clever the coding, it was not possible (without manual intervention) to arrange for embedded sub-routines, or to skip certain lines of code, or to branch to another segment of the program; the main sequence mechanism had to read every line of code in sequence unidirectionally; reversal or backstepping of the mechanism was impossible. This was why a great deal of maneuvering of tapes involving human intervention by the operator was usually required to complete a program of any appreciable complexity. This—to me, one of the most frustrating aspects of the Mark I design—later led me to incorporate a set of "subsequence mechanisms" in Mark I, which then made possible both conditional and unconditional program branching. In the meantime, the best programming techniques called for keeping operator intervention to a minimum, both to reduce manual operational errors and to reduce run time.[4]

The fourth phase called for the detailed logical flow layout of each program tape. Here the actual proposed sequence of operations must be set forth including the build-up of partial results. I found that a dynamic storage register allocation chart was extremely helpful. The checking techniques to be incorporated had to be planned and positioned in the program. Any shifting of quantities for normalization purposes or any other purposes such as printing had to be annotated at this time. With regard to normalization, it was advantageous to keep quantities as close to unity as possible so as to avoid extensive digit loss in certain multiplication and division operations, while at the same time not allowing any overflow beyond the maximum allocation for integral places to the left of the decimal point. It was clearly important to trace the proposed computation sequence carefully, taking into account for example worst-case ranges of operands, to determine the potential loss of significant figures; otherwise the validity of the asserted precision of the final output quantities would be suspect. Strategic compensating shifts by powers of ten were often effective in catering to these requirements, and the provision for such shifts had to be planned at this time.

Checking the computation was an absolute necessity; and this subject required an extensive amount of the programmer's attention in this phase of the program planning. The sources and types of potential error were numerous, the most significant being these:

• Setup errors by the operator. These included erroneous switch settings, improper plugging, failure to follow rerun instructions, improper arrangement of punched card input decks.

• Inaccurate punching of main sequence control tapes, value tapes, functional tapes, punched card inputs, etc.

• Machine malfunctions. Included in this category are failures in the basic arithmetic operations, reset failures, switch and relay contact

4. The limited storage capacity of Mark I—essentially 72 registers of 23-decimal digit length—also directly affected the planning of instruction tapes as well as the planning of input and output. Often the number of quantities involved in the computation combined with the need for "working storage" far outstripped the capacity of the machine. This in turn made it necessary to segment the problem, leading to multiple instruction tapes as well as punched card input and output in many of the runs.

failures, short and open circuits in wiring, and electromechanical input and output device breakdowns. Certain of these were positively devilish to track down. I can recall numerous instances of *digit-sensitive* failures where, for example, one particular storage register would malfunction only when there were nines stored in the fourth and fifth columnar positions! The program could run error-free for an hour before this precise configuration might recur, at which time the failure might or *might not* recur.

• Programming errors. Coding mistakes included calling for wrong operands or wrong arithmetic operations, failure to reset storage registers, omissions, duplications, faulty interposition of instructional codes, improper shifts, and every other conceivable oversight or outright blunder that one can possibly imagine.

• Errors in the programmer's calculation of input constants or starting values.

• Incorrect operating instructions formulated by the programmer. In particular, errors in plugging instructions could wreak havoc, since subtle mistakes here could cause output results to appear "reasonable" when in fact they were erroneous.

I found the task of devising and applying techniques for detecting these many categories of error both demanding and challenging. In the final analysis, of course, verification of the accuracy of the output results was the end objective. In some cases, independent end-checks against values previously established as being accurate could be utilized. For example, in the calculation of the Bessel function J_0, I had access to previously published values of the function for certain common values of the argument. If these values were identical digit for digit with my calculated values—at least through the number of decimal digits shown in the published table—there was an extremely high probability that my computations were accurate. However, I utilized appropriate high-order differences to verify proper behavior of the function for the extremely small subintervals of the argument used in our tabular computation. It was also possible to calculate the function—laborious though it was—on a desk calculator for a few widely spaced values of the argument; if agreement with the Mark I values were obtained, then in conjunction with the differencing checks, I judged that the probability was nearly unity that all values of the function for the hundreds of intermediate argument values were also accurate.

Within the computation it was possible to check arithmetic operations by reversing operands or inverting the operation. However, in effect what is being checked by this technique is the process itself, not the result. If one of the operands in a multiplication is incorrect, then reversing multiplier and multiplicand serves no purpose; the product is identically wrong in both instances. The same reasoning shows that to check a series of required quotients having the same divisor by comparing, say, the quantity

$A/P + B/P + C/P$

with the quantity

$(A + B + C)/P$

verifies only that the dividing unit of the machine is functioning properly; an error in the quantity C and thus in the required quotient C/P will pass undetected.

Checks to verify the accuracy of the punching of initial input cards or to verify the proper manual settings of decade switches were accomplished by the simple expedient of calling for a preliminary printout of the card contents or the switch settings; the typed output was then meticulously compared by hand with the original source data from which the cards were punched or the machine switches set. The vital necessity for this precaution can be gleaned from the following example. It is required to solve n linear equations in n unknowns with the $(n^2 + n)$ coefficients in the equations stipulated. If any coefficient initially enters the machine incorrectly, then the entire solution involving the execution of possibly thousands of lines of programming is invalid. What is more disturbing is the fact that all internal checks incorporated in the program will have been satisfied perfectly. In fact, if a super overall check were programmed, comparing the results obtained by determinant calculation versus the solution via successive elimination of variables, the two final results would clearly agree identically—identically *incorrect*. Furthermore since there is no precedent solution with which to compare, the programmer is blissfully unaware that all the computation is for naught. One possibility, it may be argued, is to precompute by desk calculator n quantities which are the *sums* of the coefficients of each of the n equations; then, program the machine to perform similar sums before proceeding with the solution, the desk-calculated sums having also been entered in the machine as a cross-check. This will presumably catch an incorrect coefficient, provided two successive coefficients have not been interchanged via an inadvertent misordering of two successive punched cards which may have been used to input the coefficients, in which event even this check is invalid. It is possible, however, to have the machine check the ordering of the cards by arranging for the punching of serial numbers in the cards; then, via appropriate programming, the sequence number may be segmented and validated internally when the card is read and before the actual coefficient contained elsewhere in the card is utilized in the computation. The error possibilities and the counter-measures devised to combat them often seemed endless.

In the case of any internal check operation, it was possible to stop the machine automatically whenever the checking condition was not met; while if the condition was met the machine would proceed, executing the program without interruption. To effect this, one of the storage registers known as the "check register" was specially designed to halt operation if the absolute difference of two quantities being compared, when sent to this register, equaled or exceeded a predetermined tolerance which had been transmitted to this register; this tolerance was usually set to require the difference to be zero to avoid a cessation of operation.

The fifth phase in the programming cycle consisted of the actual coding of the program. With the aid of the charts, flow diagrams, and notes created in the previous phase, it was now time to generate the series of machine instructions which would later be punched in the main sequence control tapes and constitute the program to be executed. This program was of course written in machine or object code (to use today's terminology).[5]

In the coding process, it was essential to keep careless errors to an utter minimum. For example, failure to reset a storage register prior to use could foul the computation in a peculiar and not always

---

5. "Compilers" and "source code" were unknown concepts in this era. I should, in passing, venture the observation that even were today's compiler design techniques available in 1944, the designer would have had himself one tough assignment in concocting a reasonably efficient compiler. The "run time" of the compiled program would undoubtedly have suffered greatly as compared to a highly efficient machine-code version; and the difficulty in meeting the time requirements for completing solutions to problems especially where extensive runs were involved might well have had an adverse effect on the overall usefulness of Mark I, even though there were few alternative means of computation in that day.

immediately discernible way. If perchance some small quantity were previously resident in this register the failure to reset would cause the variable transmitted to this register to be augmented by this small quantity—this because a transfer to a register does not erase the previous contents, but rather adds to said contents. Later, in test running the program, this type of error could be painfully time-consuming to isolate.

I tried to discipline myself in the coding process to "get it right" the first time. I carefully annotated the code using mathematical symbolism pertinent to the problem being solved. I marked the quantities being transferred as well as the location of partial results in order to assist in tracing the flow of the program, and I maintained a dynamic series of "snapshots" of the storage register contents as the program progressed. Figure 1 shows an actual page of my original programming for the tabulation of the Bessel functions, where my annotations can be clearly seen to the left of the lines of machine code; at this stage of the program, differencing operations were in progress.

Once the coding process was completed, and before any test running on the machine was attempted, the programmer had to proceed with the sixth phase, namely the preparation of operating instructions. Here, several areas had to be addressed. Plugging charts were drawn up for all plugboards which were to be active in the run. These boards were very similar to the plugboards utilized at that time to control data transfer in various forms of IBM card equipment such as printing tabulators. Plug wires of several types and lengths were to be inserted in these boards to control a host of functions. The instructions for plugging took one of two forms: a) explicit instructions in writing defining the plugwire array or b) printed diagrams of each board with the desired wiring drawn in. An indication of the extensive role that plugboards played in Mark I can be gleaned from a listing of the printer functions that had to be properly plugged on the printer's plugboard. These functions included printing machine decimal columns, printing and positioning of the decimal point, printing a minus sign, dropping off zeros to the left of integral portion, dropping off non-significant zeros to the right of the decimal portion, setting spaces vertically and horizontally, controlling tab positions and carriage returns, and adding a "half-correction" to yield properly rounded off output correct to a given number of decimal places. Other units to be plugged included the card feeds, the card punch, the multiply and divide units, interpolators, and many special shifting registers. The

⑦ PROB   P   TAPE. 71

	Out	In	Misc.
	765432		7
$\cdots - y^4 \to$ (or) $y_9 - y_8 + y_3 - y_4 \to 541$	6541	541	732
$\cdots + [y_9 - y_8 + y_3 - y_4] \to$ (or) $\boxed{\Delta^6 y_3} \to 6$	541	6	7
	5431	5432	7
$-\{\Delta^4 y_4 - \Delta^4 y_3\} = \Delta^5 y_3 \to 5432$	543	5432	732
$\underline{240}$	6	61	
$0.0225 [\Delta^2 y_4 + \Delta^2 y_3] \to \begin{cases} 641 \\ 64 \end{cases}$		641	7
or $_4\underline{P-2}$	86	64	7
	541	761	732
$\{\Delta^6 y_3 - \Delta^6 y_2\} \to$ or $\boxed{\Delta^7 y_2} \to 61$	54321	61	732
$\Delta^4 y_4 + \Delta^4 y_3 \to 5431$	543	5431	
	75421		7
$\Delta^6 y_3 + \Delta^6 y_2 \to 6$	54321	6	
			7
$\cdots + 7 [y_8 - y_9 + y_4 - y_3] \to$ or $\boxed{\Delta^8 y_2} \to 621$	86	621	7
$\underline{250}$	5421	761	7
$\Delta^8 y_2 + \Delta^8 y_1 \to 621$	62	621	7
	6431	6431	
	7654321		7
	643	643	7
	64321	64321	7
	71	71	7
			7
$0.006 \cdot \Delta^3 y_4$ $\begin{cases} 6421 \\ 642 \end{cases}$	8421	6421	7
(or) $_4\underline{P-3}$	86	642	7
$\underline{260}$	5431	761	7
Print Arg, on #2	2	74321	
+ A.C.	87	75221	
	8		
	6432	6432	7
	651	651	7
	65	65	7
	6521	6521	7.
	652	652	
$0.00391875 \cdot [\Delta^4 y_4 + \Delta^4 y_3]$		6431	7
(or) $_4\underline{P-4}$ $\begin{cases} 6431 \\ 643 \end{cases} \underline{270}$	86	643	7
	5432	761	7
	6531	6531	7
	653	653	

Figure 1
Programming for tabulation of Bessel functions.

transcendental functions, if used in the program, also required plugging. Since the plugging arrays on many of these boards varied significantly with the particular program tape being run, great care had to be taken both in the instructions and the actual plugging of the board to avoid costly time-consuming errors.

Settings for all switches had to stipulated, including the data switch-bank used to introduce constants into the program, as well as numerous control switches governing the operation of various machine units.

An essential part of the operating instructions dealt with actions to be taken in case of an error or "check stop." Rerun instructions were explicitly detailed, and the actions required often varied with the position at which the program stopped; frequently reruns were to be initiated at different lines of code in the main sequence tape. The programmer will have arranged matters such that most, if not all, of the useful computation preceding the stoppage will be salvaged and the rerun will seamlessly pick up the computation with a minimum of overlap. If the error persists, the operator may be instructed to read out of certain storage registers whose contents may provide an immediate clue to the trouble.

Even under trouble-free running conditions, there were usually extensive activities required of the operator. At times he was required to change switch settings after a certain number of tape revolutions. In other cases, certain actions were required if an output variable reached a particular threshold value. Routinely tapes had to be changed, decks of punched cards assembled and verified for input, output decks verified, and printer activity monitored. If the printing for example was to be photocopied for subsequent publication, such as was the case in the computation of the tables of Bessel functions, the operator had to verify that the print quality met specifications, and that the columnization and spacing of each page was flawless. Figure 2, which reproduces a segment of the actual operating instructions I prepared for the running of "Problem E" for the Navy in June 1944, shows the extent of detail typically encompassed in these instructions.

The seventh and final phase of the programming cycle involved the test running of the program on Mark I. The switches were set, the plugging completed, the input and output units activated, and the main sequence tape, which had been punched by a separate unit in accordance with the written machine code, placed on the main control mechanism. Although there were some surprising first-time successes, the usual scenario was perhaps a half-minute of running before the machine stopped on a check. It was now necessary to obtain a complete readout of all pertinent storage positions, so that the nervous programmer—who was always in attendance at this event—could see whether this printout bore any resemblance to the values that the registers were expected to contain. Of course, to make matters more interesting, Howard Aiken was usually nearby, which tended to heighten the tension. Even though Howard recognized all too well the necessity of this debugging procedure, he insisted that most of the programming errors should have been eradicated by a highly competent programmer before any trial run on the machine. Aiken expected a certain amount of unproductive machine activity, but since debugging activity was actually interrupting certain high priority programs that were in progress and already running on Mark I at full throttle, he had very little patience for an error-infested trial session. Needless to say, procedures were developed that effectively minimized the time required for the "program purification" process. Voluminous printouts were taken to the programmer's desk, where the evening hours would be spent re-analyzing the program in preparation for another trial at 3 o'clock the next morning, when the density of onlookers would be at a minimum.

Three Noteworthy Programs

One of the first problems from the Navy for solution at the Harvard Computation Laboratory ("Problem E") dealt with ballistic analyses for the 5-inch 38-caliber anti-aircraft gun. To begin with, a set of curves was furnished us dealing with several key input variables which were to be the basis of later fire control calculations. In essence, a family of curves had been developed, each curve of the family being a graph of $V = f(R)$ for a particular value of still another variable E. Previously, in order to determine the value of V for any specific values of R and E, a laborious manual graphical interpolation in *two* dimensions had been required to approximate $V_k = f(R_k, E_k)$. Upon careful study of these curves, we determined that it would be feasible to fit a polynomial of the form

$$V = aR + bR^2 + cR^3 + dR^4 + eR^5$$

to the family of curves by representing each of the coefficients a, b, c, . . . in turn in the form

- 1 -

<u>TO   OPERATOR</u>

<u>Operating Instructions for Problem E</u>

Tapes Used: Sequence: <u>9-A; 9; 10-A-ST; 10-A</u>

                         <u>Interpolation:</u> Sine Tapes: 17-A; 17-B; 17-C

                         Zero Order: 20

The Input Data: Six Decks of Punched Cards.

The first deck will be numbered: 1, 2, 3, 4, --- up to N.

The second deck will be numbered: 1001, 1002, 1003,---up to 1000 + N.

The third deck will be numbered: 2001, 2002, 2003,---up to 2000 + N.

The fourth deck will be numbered: 3001, 3002, 3003,---up to 3000 + N.

The fifth deck will be numbered: 4001, 4002, 4003,---up to 4000 + N.

The sixth deck will be numbered: 5001, 5002, 5003,---up to 5000 + N.

Run 1:   <u>Rotation of axes.</u>

       1. Place Sine Tape 17-A on Interpolator No. 1
            "   "   "   17-B   "   "   No. 2
            "   "   "   17-C   "   "   No. 3

    WARNING: Make certain that the imbedded pins on both sides
              of the drum are in line with printed line numbers
              on the tape.

       2. Place Tape 9-A on sequence mechanism.

       3. Twelve cards are placed in Card Feed No. 1.
           The cards are fed and thus arranged in this
           order: 4001, 3001, 5001; 4002, 3002, 5002;
           4003, 3003, 5003; 4004, 3004, 5004.

       4. Decimal Point lies between <u>9 and 10.</u>

       5. Switches: Set switch No. 31 to 1.0
                        "   "   No. 32 to 0.5
                        "   "   No. 33 to 1.57079633

       6. <u>Typewriters</u> are not used.

       7. <u>Plug Card Feeds</u> for <u>direct</u> reading into machine.

       8. <u>Multiplying</u> Unit plugged for 9 decimal places.

       9. <u>Division</u> plugged to 12th place.

Figure 2
Operating instructions.

- 2 -

Run 1: (cont.)

10. Interpolator Plugging:

  Highest Order h-column is 8.
  Plug "C" values direct.
  All three interpolators are
  plugged identically.

11. Switches above all three interpolators are set as
follows:

  One-half no. of "A" values = 33.
  No. of "C" values = 4.
  (Set both rotary switches to four.)

12. Special Switches:

  Set "Divide N-decimal" to 13.
  Set "Log N-decimal" to 13.

13. Logarithm plugging: Plug for negative logarithms.

14. Exponential plugging: Plug for decimal between 9 to 10.

15. When red light signal shows cards out in
feed No. 1, flip switch on feed to "off"
position, push start key and allow sequence
tape to go on to its very end. (Will end without depressing any stop keys.)

16. Put on tape to read out of all counters
successively.

17. Plug typewriters No. 1 and No. 2, for decimal
point between 9 to 10.

18. Plug typewriters No. 1 and No. 2 for vertical
spacing as follows:

  No. 1: Groups of three.
  No. 2: Groups of two.

19. Flip on typewriter No. 1 and proceed to read out
of all 72 counters and print.

20. Bring this printed extract to the coder of the
problem for final check and examination.

21. Tape No. 9 is now put on the sequence mechanism and
placed on an 87 code at "Start".

22. The remaining cards of the 4-thousand; 5-thousand;
and 3-thousand group are placed in feed No. 1 in the
same sequence as before.

  The twelve cards of the previous run are not to be
  repeated in this run.

Figure 2, continued

- 3 -

Run 1 (concl.)

        22. (cont.)

            The arrangement of cards for this run
            should be: 4005, 3005, 5005; 4006,
            3006, 5006; 4007, 3007, 5007; etc.
            through to the end.

NOTE:

    In case a run using Tape No. 9 is interrupted or an emergency stop
is used, the following procedure is to be followed:

        1.   Go back 5 groups of 3 cards each --- for example:
            suppose a run to be interrupted where the last
            cards fed were: 4091, 3091, 5091. Reform the
            deck carefully with the top card numbered; 4087
            and the cards following being 3087, 5087; 4088,
            3088, 5088, etc.

            Now take the top 12 cards from the deck as it stands.
            In the example above, this would mean cards 4087
            through 5090 inclusive. These cards are placed in
            Feed No. 1 and the run using Tape No. 9-A is per-
            formed as detailed in steps 1--20. When this is
            done, go on to Step 21 and proceed with the run.

        2.   Tapes 9-A and 9 require exactly the same plugging
            throughout.

        3.   The typewriters are plugged as specified in Steps
            17 and 18.

        4.   Make certain that both typewriters are on before
            beginning the run with Sequence Tape No. 9.

        5.   "Print Complement" switch should be on for Type-
            writer No. 2.

        6.   All switches and interpolator arrangements are set
            exactly the same with Tape No. 9 as with Tape No. 9-A
            (Consult previous pages of the instructions.

        7.   The run will end after the Typewriter No. 1 has
            printed straight zeros 13 successive times.

        8.   A short time before the end of the run, cards will run
            out on Feed No. 1. Flip the switch to the "off" position
            and continue to run until Step 5 (immediately above)
            has been carried out.

        9.   Time per single rotation of tape = 11 min. 20 sec.
            Time per group evaluation = 2 min. 52 sec.

Figure 2, continued

$$a = k_0 + k_1E + k_2E^2 + k_3E^3 + k_4E^4 + k_5E^5,$$

where the k_i are a different set of constants for each of the coefficients. This was accomplished by first fitting a quintic polynomial to each of six of the V curves (V plotted against R for a fixed E); the resulting six values of each of the coefficients a, b, c, . . . were then used to fit a separate quintic polynomial in the second independent variable E to each of the five sets of six coefficient values. The result was a "biquintic" (i.e. a quintic in R whose coefficients are a quintic in E) polynomial expression for $V = f(R,E)$ with a total of 30 derived constant coefficients. This entire process was repeated to curve fit a biquintic to another graphically represented variable, namely $T = g(R,E)$. Finally still a third function $D = h(T,E)$ was curve fitted, this time using a quadratic in T with coefficients which in turn were quadratic in E.

These extensive polynomials proved to fit the data with minimal observed error, and thus enabled us to evaluate each of the three key variables via an explicit expression and without recourse to interpolation of any kind. The main ballistic calculations then proceeded utilizing several other input parameters in addition to R and E, including various angular, time, and distance inputs. The entire program, including the curve fitting, involved some fifteen different main sequence control tapes, as well as several interpolation tapes used to determine values for circular functions required in the computations. It is of interest to note that the polynomial functional representations discussed above became the subject matter of the laboratory's first official report dated June, 1944: *Report No. 1* of the *Computation Project of the Bureau of Ships* (the name given by the Navy to the Harvard Laboratory).

In the first days of August 1944 a problem was brought to the Laboratory for solution on Mark I by the renowned mathematician and physicist John von Neumann. The problem involved solving a nonlinear partial differential equation of the second order. The equation described a spherically symmetric flow of a compressible fluid in the presence of a spherical detonation wave proceeding inward from the surface of the sphere toward its center. I was to work with von Neumann and two of his associates—Valentine Bargmann and Charles Loewner, mathematicians affiliated with Brown University at the time—in organizing this problem for solution on the machine. Any attempt to solve this equation through some technique using successive approximations would lead to results whose accuracy could not be determined let alone relied upon. The only recourse was to effect a direct numerical approach using difference-equation techniques. Accordingly, $y = F(x,t)$ was to be determined numerically by creating a fine mesh in the two-dimensional x-t plane across which a numerical solution for y and certain other variables important to the physical problem would be effected—mesh point by mesh point—within the pertinent domain. The mesh was traversed in the x direction (the radial space variable) from the surface toward the center—i.e. for successively decreasing values of x—for each value of the time variable t beginning with the value $t = 0$. The computation was to continue until a certain critical value of pressure p was reached within the sphere, this being a derived variable computed as the solution at each mesh point proceeded. Further complicating matters were certain discontinuous initial boundary conditions requiring a special method for obtaining a first approximation of certain beginning values near $t = 0$ in order that the difference equations could then be applied in the main process. Also, cubic and quadratic interpolation was required at the boundary with the shock wave front before the computation could proceed to the next time increment. Once the governing boundary conditions and difference equations were specified by the team of mathematicians, the programming of the problem became my responsibility.

It was necessary to program several sequence tapes yielding certain required variables which were to be used repeatedly in the main run. The objective here was to enable the main run to proceed without interruption. This main run consisted of evaluating the required quantities at each "x-step" for a given "t-value"; these quantities were punched into cards at each x-step to be used later as inputs when the main run was to be repeated for the next value of t—the time variable. Since special computations and interpolations were required as x reached the inner boundary of the domain at each t-step, manual tape maneuvers were necessary at that juncture.

Once the programming was completed and debugged, machine runs proceeded. Unfortunately, the results that came forth from these initial runs indicated that the mesh being used was too coarse; accordingly a much finer mesh was instituted in which the radial or "x" dimension consisted of 200 units and the time dimension of 240 steps. This latter arrangement produced what von Neumann considered a satisfactory solution to the problem; the program ran steadily through much of the month of September 1944, yielding an enormous quantity of data as a result of the thousands of points constituting the mesh.

My natural sense of curiosity was aroused when the problem first arrived on the scene. First , the fact that von Neumann was personally involved suggested immediately that this was of great importance. Second, I did know that there were highly secret operations related to the military efforts in progress at the Los Alamos National Laboratory, and, since I knew that this problem had emanated from Los Alamos, I knew it had to be urgent in nature. In proper Navy tradition, I did not probe further nor discuss it further with anyone, but there was no question in my mind that this dealt with an explosive device with some special detonation properties!

Although during the period of preparation and while the problem was being run, John von Neumann showed a degree of curiosity with regard to the design and programming characteristics of Mark I, I recall that we spent very little time going into much detail on these matters. In fact, he confessed that the computer area was pretty much a new subject for him. Most of our discussions were confined to the problem at hand and the best means for solution. I can't help but wonder, however, whether the complex programming and operational manipulations which he witnessed, including the many instruction-tape changes that were required to solve the problem—whether all of this was not duly noted by von Neumann and may have had some bearing on his contributions in the field of computer design several years later.

Details on the solution of this nonlinear partial differential equation, including the numerical methods and difference equations used, as well as tables of results in the x-t plane were incorporated in Report No. 11 of the Computation Project of the Bureau of Ships issued at Cambridge, Mass. in December 1944; the document was classified Confidential. Over my persistent objections, von Neumann generously insisted that I be shown as the prime signatory on the document, maintaining that in reality I had done most of the work, and the report was thus issued.

At the request of the Naval Research Laboratory in late 1944, the Navy's Bureau of Ships authorized the Harvard Computation Project to compute and tabulate extensive tables of the Bessel functions of the first kind—$J_n(x)$—from order (i.e. $n = $) 0 through 100. Such tables did not exist to the accuracy and for the arguments desired, and several advanced research projects being pursued by the Navy had need for these functional values. Late in 1944, I was directed by Howard Aiken to commence the programming for the production of these tables. Before any work could begin, extensive analytical sessions were held at the Laboratory dealing with formulae to be used in the evaluation of the functions, accuracy requirements, argument intervals, and verification of results. It was eventually decided that J_0 and J_1 were to be computed with sufficient accuracy to yield tabular results correct to eighteen places of decimals. The arguments for both of these functions would proceed from $x = 0$ to $x = 25$ by intervals of 0.001; from $x = 25$ to $x = 100$ the interval would be 0.01. The functions J_2 and J_3 were also to be computed with sufficient accuracy to yield the same eighteen place precision for similar intervals of the argument. The higher-order functions J_4 through J_{100} were to be computed so as to yield a precision of ten decimal places in the final tabulation.

It became immediately clear that for the accuracy desired even the 23 decimal places of Mark I were insufficient! The machine had to be converted to a 46 decimal place machine to accommodate the successive effect of round-off or truncation errors inherent in the arithmetic operations to be performed. Accordingly, the chief resident engineer, Bob Hawkins was requested to "gang" two pairs of storage registers so as to create effectively two 46 decimal place registers; the 23rd column of the low-order register would carry to the first column of the high-order register while the 46th column (the 23rd column of the high-order register) would carry back to the first column of the low-order register. These two register pairs were termed double-precision registers and were callable by a set of new instruction codes. Thus addition of double precision numbers could be effected. All that remained was to effect double precision multiplication, since the machine's division instruction was not to be used in the computations. Fortunately, double precision multiply required no reengineering of Mark I. The product register in a standard multiplication already provided a full 46-digit product; therefore the double precision product $(a + b)(c + d)$ could be effected by the addition with appropriate shifting of three products, namely $ac + bc + ad$, the additions being performed in the double precision registers. The product bd of course is ignored, having a value of less than 1 in the 46th decimal place. All that was required was a special instruction code to transfer the low-order 23 digits of the product register to the main buss.

Various evaluation techniques were used to keep the computation time to a minimum, while maintaining the high degree of precision required. In the argument range for x less than 25 the ascending power series for $J_n(x)$ was used for the low-order functions. For small values of x, twelve terms of the series yielded the necessary convergence; for larger values the number of terms increased, with as many as 57 terms of the series being required as x approached 25. The coefficients in the series contained reciprocal factorials and were precomputed by desk calculators to 51 significant figures to preclude— with something to spare—any cumulative roundoff errors from affecting the 23rd decimal place of the final result. Any errors in these coefficients would negate untold hours of machine operation before the error were discovered; accordingly I personally performed the arduous labor of making these calculations on a Friden desk calculator, then checking and rechecking the results. The task of performing divisions accurate to 51 places on a ten-decimal digit desk machine could have been termed cruel and unusual punishment!

Subtabulation using the Newton-Bessel central difference formula was performed in the range $x = 2$ to $x = 25$ for the functions J_0, J_1, and J_2; this subtabulation to the interval 0.001 was based on the differences derived from the power series values at interval 0.01.

A third technique involving the asymptotic expansions for the three lowest-order functions was invoked in the range $x = 25$ to $x = 100$ since the convergence of the ascending power series in this argument range would have required a prohibitive number of terms. The desired convergence was obtained with 23 terms of the asymptotic series at the lower part of the argument range and as few as twelve terms in the higher argument range. Coefficients for these series were, as before, precomputed on desk calculators in this instance to thirty-one significant figures. The requisite sine and cosine functions, as well as the inverse square root were all calculated on Mark I in special preparatory runs—all of these functions appearing in the asymptotic series.

Finally, a fourth mode of evaluation—and the simplest to perform— was based on the recurrence relationship that holds for three successive orders of the Bessel function. This formula,

$$J_{n+1}(x) = (2n/x) J_n(x) - J_{n-1}(x),$$

was used to obtain the higher-order values of $J_n(x)$, including all orders $n = 4$ through $n = 100$; recurrence was also used as an independent check to verify all values of $J_0(x)$, $J_1(x)$, and $J_2(x)$ which had previously been computed by the ascending power series and the asymptotic series.

Even with the high decimal place precision, in order that as few significant figures as possible would be lost in the repeated multiplication operations, it was mandatory to use normalization techniques in the computation of all series expansions. This was necessary so as to keep both the coefficients as well as powers of x in numerical balance, the latter normalized so as to be close to but less than unity.

High-order differencing was used extensively to check the validity of the results. Usually the final differences were less than one unit in a decimal place three columns beyond the last printed decimal place of the final tabulation. The smoothness of these high-order differences together with "spot" verification by alternate computation at pivotal argument intervals gave us a high degree of confidence in the absolute accuracy of the results. These differences were calculated from tabular values of the function just before the round-off operation in preparation for printing; a separate validity check on the round-off process was also incorporated in the program.

All printing of the Bessel function tables was done on the IBM typewriters connected to Mark I; these machines were specially equipped with rolls of paper carbon ribbons to give as fine a definition and contrast as possible in the printed output. It had been determined at the outset that there would be no human processing or transcription of the printed results, and that the tables to be published were not to be typeset, but would consist exclusively of photo-offset reproduction of the actual original printed output from Mark I. The clarity and resolution of the printed tables were exceptional, and as of today, a half-century later, the caliber of print remains unmatched—even by the most recent laser printers—for direct computer printout.

During most of 1945 and well through 1946, the Bessel-function tabulations remained a base load problem, running continually on Mark I unless problems of higher urgency intervened. In terms of programming, no other problem quite approached the magnitude of the Bessel-function tabulation; I have counted more than eighty main sequence tapes, the programming for which still reside in their original loose-leaf notebooks in my library. The final tabulations were published in several volumes of the Annals of the Computation Laboratory of Harvard University, beginning with volume III in 1947.

A Major Design Modification

Ever since I had begun to program for Mark I, one major disadvantage inherent in the machine's design repeatedly reasserted itself. Unfortunately, no provision had been made for any variation in the fixed sequence of instructions governing the machine's operation; the broad-width sequence tape in which successive instructions were punched was read by the sequence mechanism unidirectionally one line of programming at a time without exception from beginning to end. The best the machine could do automatically was to execute what I had termed at the time a "one-legged branch." This was accomplished through the use of the "check register"—storage register 72 with specially designed controls for this purpose. If one side of an inequality prevailed, the machine would continue on with the next instruction; if the other side prevailed, the machine would come to a grinding halt—its only available alternative. Once the machine stopped, the operator, alerted to this situation, could then manually advance or reverse the main sequence tape to some new position in the program. When an unknown number of iterations of a process was required, this crude time-consuming procedure was once again the only method of coping with the situation. The von Neumann problem in mid 1944 was a particularly frustrating example of this major flaw in the Mark I design; I was forced to resort to several manual operator interventions in the program's sequencing to execute the requisite branching. At that time I had already come to the conclusion that something had to be done to effect automatic branching in Mark I, and between programming assignments I began to think about and sketch out possible design changes for accomplishing this feature. Unfortunately, this was not a time to consider major changes or modifications to the machine and the subject was shelved until mid 1946, shortly after my discharge from the Naval Reserve, at which time I continued my work at the Harvard laboratory in a civilian capacity. In late August of 1946 I set forth what seemed to be a feasible solution for incorporating a full branching capability in Mark I; I discussed the plan with Howard Aiken, who, as usual gave his undivided attention to the subject; his suggestions and critical review proved invaluable.

The plan was to create a self-standing "sub-sequence" unit to be connected electrically to the main machine; this unit was comprised of ten relay operated stepping switches each capable of providing 22 lines of program code. The content of each of the 220 lines of code was determined by a plugboard panel in which any possible code combination in the OUT, IN, and MISC sections was called forth by appropriate plugging. These lines of programming were identical logically to the lines of programming which were punched in the main sequence tape; and they governed the operation of Mark I in the very same way. When conventional instructions were read by one of the subsequence steppers, each line was read in sequence, with the device stepping to the next line and thence reading the new line under control of either the internal "automatics" or a "7" in the MISC column of the previous line of code—much in the same way that the main sequence mechanism had always operated.

Here however the similarity ended; new instruction codes were introduced which enabled the ten sub-sequence steppers (SS#1–SS#10) to initiate actions program-wise which were never before possible. In particular, a series of codes provided for an *unconditional* or absolute "call," causing the indicated SS unit or, alternatively, the main sequence mechanism to take over control; the unit thus called would now govern the next instruction to be executed by the machine. A further set of codes provided for a *conditional* call causing the sequence unit indicated to assume instructional control only if the algebraic sign of the quantity located in storage register 70 was negative; if the sign was positive then the conditional call does not take effect, and the sequence unit currently in control simply proceeds to its next line of coding without surrendering control. A further array of codes enabled the programmer to step any SS unit forward one line as well as step the main sequence mechanism one line forward without reading—the latter feature not previously possible. Finally, codes were assigned which caused a reset of any SS unit from its then current position to its first line of coding.

With the installation of this new subsidiary sequence capability, a broad degree of flexibility in the instructional sequence was now possible. The programmer could now perform multi-level branching operations as well as a variable or fixed number of repetitions of subroutines or iterative procedures without any interruption of machine operation and with no necessity for the awkward manual resetting of the main sequence mechanism. Furthermore, since the coding was pluggable, it was possible to make changes or correct errors in the program without having to repunch or patch an instruction tape. A panel of the new unit was also equipped with indicator lights designating the current line position of the active SS unit—an aid in any debugging or error-tracing process.

The design of this sub-sequence unit was an exciting and challenging experience for me, and gave me a good deal of personal satisfaction. The unit clearly had to be consistent with the main machine with regard to timing, relay circuit design techniques, and overall logic; it was also essential that the new unit not compromise or modify any of the standard operations of the original machine; and finally, all program tapes contained in the large Mark I library, which had grown to substantial size by this time, were to run without any modification whatsoever. Furthermore, computer-aided design was not yet born, and any gross errors which were not ferreted out by a careful deskside review of the circuit design and the timing charts could weave their way into the wiring and relay connections with devastating effect. Not only would the unit have to be torn apart and reworked, but the operational fidelity of the main machine could well have been affected in the process.

By the time the Subsidiary Sequence Unit was built and installed in Mark I, I was about to become an alumnus of the Harvard Laboratory, having received an offer I couldn't refuse—to enter the commercial computer world; this was now March of 1947. Consequently I was not able to participate in the most exciting moment of all—watching the power being turned on and anxiously witnessing the birth throes of my creation; nor, regretfully, did I ever have the opportunity to compose a program utilizing the new feature. I never did determine whether my "baby" gurgled happily or threw a temper tantrum upon its debut, but I was told that the unit became a sine qua non in almost all subsequent programs that were coded for Mark I throughout its long and productive life at Harvard, and in that knowledge I am gratified.

Mark II, an Improved Mark I
Robert Campbell

After the dedication of the Mark I automatic electromechanical calculator in August of 1944, and the increasingly productive operation of the calculator by the end of the summer—operating three shifts per day, seven days per week—Howard Aiken began to think about developing a successor machine. From the beginning this machine was designated Mark II. It was planned, designed, developed, constructed, and tested by the Harvard Computation Laboratory, and then delivered to the Navy in early 1948.

Although the Mark I calculator, which was being operated as a project of the US Navy Bureau of Ships, was available to do work for all elements of the Navy, as well as for other branches of the services, the Naval Proving Ground at Dahlgren, Virginia. felt the need for a machine that could be dedicated full-time to proving-ground applications. A dominant application was the solution of simultaneous total differential equations for the construction of firing tables—i.e., the tabulation of the range and flight time of artillery shells as functions of gun elevation angle, propelling charge, shell characteristics, atmospheric conditions, and other parameters. Other major requirements during World War II were for bombing tables and rocket ballistic tables, which required numerical processes similar to those used for firing tables.

During the early part of World War II, Dahlgren had supplemented its mechanical desk calculators with IBM punched-card equipment. But it was realized that more automatic computing equipment was required. Captain Charles C. Bramble and Commander Willard E. Bleick of the US Naval Reserve held early discussions with Aiken in the fall of 1944 concerning their computing needs. They discussed general technical requirements and set priorities and target time schedules. They needed a machine to be available at the earliest possible date.

At the time when the planning for Mark II began at Harvard, the development of automatic computer technology was in an early stage. Mark I had been designed and built using Aiken's system concept and IBM electromechanical components—dominantly electrically controlled mechanical decimal "counters" for number storage and addition, and electromechanical relays for control. The only new elements developed particularly for Mark I were the punched "paper" tape units for input of instructions to implement Aiken's concept of "automatic sequence control" and for input of sets of numbers and tables of functions. The medium used for the tape was uncut IBM punched-card stock.

The Bell Telephone Laboratories had developed and placed in operation a succession of increasingly versatile and complex machines using electromechanical relays for number storage, addition, and control. The Model 5 machine, brought into operation in 1944, was a general-purpose machine somewhat similar in overall capability to the Harvard Mark I. It used Teletype printing telegraph equipment with punched paper tape for input and output.

The first electronic digital computer, the ENIAC, was under construction at the University of Pennsylvania, for subsequent delivery to the Army's Aberdeen Proving Ground. Aberdeen had computing requirements similar to those of Dahlgren, and the ENIAC design had emphasized the computation of trajectories, although the machine had general-purpose capabilities. While changing problems on the ENIAC was initially quite cumbersome, later enhancements to the machine made changing problems much easier. Built using vacuum tube decimal ring counters for number storage and addition, ENIAC was some thousand times as fast as the Harvard Mark I or the BTL Model 5, as far as internal operations were concerned. ENIAC would be brought into initial test operation at the University of Pennsylvania in 1945.

The IBM technology and components used in the construction of Mark I were not considered suitable for Mark II for several reasons. From the practical standpoint, a joint Harvard-IBM effort would have been very difficult, if not impossible, to achieve because of the breach between IBM and Harvard, and especially between IBM's president Thomas Watson and Howard Aiken, stemming from the time of dedication of Mark I. Aiken, a proud man, could not imagine going to IBM, hat in hand, to ask for a renewed form of collaboration.

I think that collaboration with IBM, even if it could have been arranged, was felt to be undesirable for another reason. Aiken was uneasy, I believe, with the fact that he had been so dependent upon IBM for the technology of Mark I, and felt that he needed to demonstrate that he could design and build a machine on his own. Moreover the electromechanical decimal counters upon which Mark I was based determined that the addition cycle time would be 0.3 second, resulting in a slower machine than one that could be achieved using other electromechanical devices, such as relays. Finally, Mark I operation in the early fall of 1944 was still not very reliable, in part due to the high contact resistance of the IBM relay contacts.

After considerable discussion, both within the Computation Laboratory, and between Harvard and Dahlgren, we decided against using an electronic approach. Aiken particularly felt that a period of research and experimentation would be necessary before an electronic machine could be built with confidence. (It may be noted that we had no one on board with experience in the technology of television or of radar: these fields were to provide the principal expertise in high-speed pulse circuitry.) Thus, Mark II would use an electromechanical approach, and would provide an incremental rather than a radical improvement in performance. The new machine would be based on some components to be newly developed and some standard components available from the Western Union Telegraph Company—the latter components similar to the printing telegraph equipment available to BTL through Teletype.

Accordingly, we developed the general plan for an electromechanical machine, and documented this in a preliminary report, prepared by Aiken and Campbell, and submitted to Dahlgren in the fall of 1944. Soon after the approval of this report in February 1945, work began on the exploration of available components, and the initial phases of design. It should be noted that in April 1946 Harvard was authorized by Dahlgren to proceed with research on an electronic computer that would be called Mark III. By this time the design of Mark II was well advanced and construction had begun.

Concept of the Machine

The development of the Mark II design clearly used the Mark I design as a point of departure. Changes in general design concept were introduced for three principal reasons, namely: first, to remedy some of the limitations found in the early use of Mark I; second, to utilize effectively a different set of electromechanical components; and third, to tailor the design of Mark II toward its intended use in ballistic calculations.

Despite the changes, the capabilities and characteristics that we visualized for Mark II were similar in many respects to those of Mark I. As in Mark I, numeric words were stored and manipulated using electromechanical devices, under the control of electromechanical relays. Numeric words and instructions could be introduced via punched paper tape, and small sets of numbers—computational parameters—could be introduced via manually set dial switches. Final results were printed on electrically driven serial printers or typewriters. Also, all internal operations of the machine were conducted in a synchronous manner, being timed by electrical impulses generated by a set of cam-operated contactors.

There were, however. a number of significant differences between the two machines. First and foremost, while Mark I stored and added numbers using electrically controlled, mechanically driven, rotary decimal counters, Mark II used relays, four per decimal digit. This change permitted Mark II to carry out basic arithmetic operations considerably faster than the older machine. Also, most importantly, while Mark I was a fixed-point machine, Mark II was designed to be floating-point.

Unlike Mark I as it was originally designed, Mark II provided multiple tape readers for instructions, so that more than one sequence of instructions could be available to the machine—for example, a master program and a subroutine. While Mark I used standard numeric IBM punched cards as coded input/output media (as well as punched tapes for instructions and numeric inputs), Mark II used only punched tapes for these purposes.

The choice of built-in basic arithmetic operations differed in that Mark II did not directly perform division. The complexity of providing built-in division was considered to be unnecessary, considering its low frequency in most computing routines. Mark II had controls to assist the programmer in computing six functions: reciprocal, reciprocal square root, logarithm, exponential, cosine, and arctangent. In addition, special aids were provided for introducing tables of functions and for performing search and interpolation operations.

Each of the 72 internal storage registers in Mark I was an accumulator. This was possible because of the decimal rotary counters used for storage, and because of the fixed-point representation. In Mark II, providing more than just storage capability in the internal registers was not feasible: relay adders are vastly more complicated than relay storage units, even for fixed-point numbers, let alone for floating-point ones. Hence, Mark II was provided with separate units for addition/subtraction, completely distinct from the internal memory.

A final point of major difference between the two machines was that Mark II, unlike its predecessor, was a system containing two complete computers, which could be operated either independently on separate problems or in combination on one problem. Because of their locations as seen from the front of the system, these were called the "Left (L) machine" and the "Right (R) machine." The L and the R machines each had a complete complement of functional capabilities, and each could operate independently of the other. Thus, two different problems could be run simultaneously if each problem required no more internal storage and other capacities than one L or R machine had. But also, to handle a larger problem, one could combine the L and R machines into one larger machine, albeit with some constraints in problem programming. Dahlgren also found another use for the dual system: one machine could be run "against" the other for checking purposes, or for locating malfunctions.

Representation of Numeric Words; Impact on Addition and Multiplication

In considering a machine tailored to the firing-table type of calculation, we came to the conclusion that the type of precision provided by Mark I (up to 23 digits) was not required, but that a floating-point numeric word would be desirable. Thus, a numeric word X was stored as a pair of numbers p and n such that $X = p \cdot \exp(n)$, where $\exp(n)$ is 10 to the nth power. Here p was stored in ten decimal digits plus algebraic sign, with the decimal point after the left-hand digit. Hence, the absolute magnitude of p was in the range $1 \leqq p < 9.999999999$. The exponent n could take on values from -15 through $+15$. To give an example, in this notation the equatorial radius of the earth in meters ($X = 6,378,160$) would be written as $p = 6.378,160,000$ and $n = 6$; while its reciprocal would be written as $p = 1.567,850,000$ and $n = -7$.

Physically, a numeric word X was represented in either the storage register relays or the holes in a punched tape in binary coded form. The ten decimal digit quantity p was represented with ten sets of four relays (or holes) each, plus a relay (or hole) for the algebraic sign, requiring a total of 41 relays (or holes) in all. The four relays representing a decimal digit were assigned the usual binary coded values: 1, 2, 4, and 8. The exponent n was represented by four relays (holes) plus one additional for the algebraic sign. Thus, representation of the number X required 46 relays (holes) in all. Within the machine, numeric words were transferred by fully parallel transmission over a 46-wire bus. It should be noted that while only the digits 0 through 9 (binary combinations 0000 through 1001) were utilized for the decimal digits of p, all 16 binary combinations (0000 through 1111) were utilized for the exponent n.

The use of floating-point numbers obviously complicates the addition/subtraction process, by contrast with the fixed-point approach. Thus, to form the sum X_3 of two positive numbers X_1 and X_2, the machine needed to compute

$$p_3 \cdot \exp(n_3) = p_1 \cdot \exp(n_1) + p_2 \cdot \exp(n_2).$$

To do this the machine first needed to compare n_1 and n_2. Assume $n_1 > n_2$. The p_2 needs to be shifted n_1-n_2 columns to the right before it can be added to p_1. If the sum is less than 10, it represents p_3 directly, and $n_3 = n_1$; if not, p_3 is the sum shifted one column to the right, and $p_3 = p_1 + 1$. If X_2 is subtracted from X_1, or if X_1 and/or X_2 are negative, there are obviously additional or modified steps required.

In view of the procedures required in adding and subtracting floating-point numbers, as outlined above, an addition/subtraction unit in Mark II needed a coded decimal adder/subtractor for the p quantities, a pure binary adder/subtractor for the n quantities, plus capabilities for comparison, shifting, and sensing zeros.

In multiplication, $X_3 = X_1 \cdot X_2$, the machine needed to determine

$$p_3 \cdot \exp(n_3) = p_1 \cdot p_2 \cdot \exp(n_1 + n_2).$$

To compute $p_1 \cdot p_2$, the machine made available $1 \cdot p_1$, $2 \cdot p_1$, $3 \cdot p_1$, $4 \cdot p_1$, and $5 \cdot p_1$; then selected these according to the successive digits of p_2 (working from right to left), and combined, with shifts, the selected multiples to form the product. Note that because only five rather than nine multiples are available, a combination of addition and subtraction is necessary. Then, if $p_1 \cdot p_2$ was less than 10, the exponent n_3 would be $n_1 + n_2$, and $p_1 \cdot p_2$ would be p_3 directly. If, however, $p_1 \cdot p_2$ was 10 or greater (but necessarily less than 100) n_3 would be $n_1 + n_2 + 1$, and p_3 would be $p_1 \cdot p_2$ shifted one column to the right.

Selection of Basic Electromechanical Components

The first and most urgent problem in the development of Mark II was the selection of components for internally storing numeric words, for performing the basic arithmetic processes, and for the control of machine functions. We considered using a variety of electromechanical devices, but soon concluded that some form of relay would be the most appropriate choice for all of these functions. After establishing specifications, we contacted many manufacturers of relays, and tested a number of samples, but were unable to find any existing product that met all of our requirements.

The Autocall Company of Shelby, Ohio, expressed an interest in helping us to develop a suitable relay, and Harold Seaton, an Autocall engineer, worked closely with us in the design. We were also able to obtain advice from Electrical Engineering Professor Reinhold Rudenberg at Harvard. To study prototype units, we built equipment to test speed of operation, and to determine endurance under simulated operating conditions.

Three different relay designs were developed to meet our functional requirements. All had six double-throw contacts, and were largely assembled from the same basic parts. One type had a single coil and a spring return, a second had double coils and a spring return, while the third also had two coils but was mechanically self-latching. The first type of relay would stay in the picked-up or energized position only as long as the pickup current continued to be applied to the coil. The second type could be maintained in the picked-up position by current applied to the second coil, after current had been removed from the first coil. The third relay type could be maintained in either position indefinitely by the mechanical latches, without any sustaining current having to be applied. Also each of the coils could transfer the contacts in one direction only. This relay type was utilized in the internal storage registers to reduce power consumption requirements. It also enabled the registers to retain their information even after all power had been shut down.

All the relays were jack-connected, so that they could easily be removed for testing or maintenance. All soldered connections were therefore made to the sockets, not to the relays. Mark II utilized about 13,000 relays in all. The relays were tested to operate in either direction in less than 10 milliseconds (11 milliseconds for latch relays). They were operated by control current pulses of somewhat less than $16^2/3$

milliseconds duration, drawing about 60 milliamperes at 100 volts DC. Like Mark I, Mark II was synchronous in operation: the current pulses energizing the relay coils were provided by cam-operated contactors. In general the relay contacts were used to position circuits only; circuits were made and broken by the contactors. To ensure low-resistance contacts for the relays, the contact material was silver and a minimum contact pressure of 30 grams was provided in both the normally open and normally closed positions.

For data and instruction input, and for data output, Mark II employed a variety of devices; some of these were minor modifications of standard equipment, while others were custom designed at Harvard. We decided to use paper tape for all coded input and output. The storage media utilized were standard Teletype punched tapes, using the five-hole width for numeric data and the six-hole width for instructions.

For reading numeric words into the machine in preassigned sequence, and recording coded numeric outputs, standard Teletype five-hole tape readers and punches were employed. These read or recorded two numeric words in 3 seconds, reading or punching one row at a time. Higher-speed reading devices, custom designed at Harvard, were utilized to introduce coded tables of functions, and to read instruction tapes.

For providing numeric output data in printed form, standard serial Teletype page printers were used. These provided output at the rate of about six characters per second. In addition to the on-line input/output devices that have just been described, there were also two auxiliary off-line tape devices. These were used for the manual preparation by keyboard of the numeric and the instruction tapes, and for the automatic duplication of these media.

Two other basic devices need to be mentioned. Like Mark I, Mark II utilized plugboards so that problem-dependent connections could be established between the 46-wire bus used for internal transfer of numbers, and some of the numeric input/output devices. Also, sets of manually set dial switches were provided so that a limited number of numeric constants or parameters could be introduced directly into the machine without having first to be punched into tape.

Complement of Equipment

The L and R machines each had the following complement of equipment for internal operations:

relay storage registers for 48 numeric words

one add/subtract unit

two multipliers

built-in controls for computing the reciprocal, reciprocal square root, logarithm, exponential, cosine, and arctangent

controls for the built-in table search and interpolation processes

five special functional registers

Each machine also had these input/output units:

two high-speed reading mechanisms for punched instruction tapes

two slow reading mechanisms for numeric tapes

two functional (numeric) tape readers with automatic tape positioning and higher-speed reading

two output tape punches for numeric data

two teleprinters

twelve sets of dial switches for manually inputting numeric constants

Readers familiar with today's computing machines may be surprised that the internal memory capacity of the L or R machine was only 48 numeric words. But we must remember that this was before the time of bulk memory techniques such as magnetic drums, acoustic delay lines, and electrostatic storage tubes. Without such bulk techniques we were forced to utilize 46 relays per numeric word, making large memory capacities not feasible. (Magnetic drums would be utilized in the next machine, Mark III, which was able to store internally more than 4000 data words.)

It should be noted that the allocation of one adder/subtractor and two multipliers to the L and the R machine was based upon an analysis of typical computational procedures, as will be described in the section on Instruction Format and Programming.

Physical Layout

Mark II was physically a very large machine, which required some 4000 square feet of floor space. This was more than twice as much area as that required for Mark I. The additional space requirements were due to two main factors; first, the substitution of relays—four per decimal digit—for decimal rotary counters (which latter devices added as well as stored); and second, the incorporation in Mark II of two complete computers. It should also be noted that the Mark II relays were larger and more ruggedly constructed than those of Mark I.

The Mark II layout included a long front panel, in the shape of an opened-up U; this contained most of the peripheral (input/output) equipment and the controls for both the R and L machines. The four page printers, however, were mounted on a console separated from the front panel. Behind this panel, and at right angles to its central section, were six long, narrow walk-in relay units with associated wiring. Each of the relay units had relay racks on both of its sides, with the relays plugged in on the outside and the wiring on the inside.

At Harvard, the machine was assembled in the old Gordon McKay building (which has long since been torn down). This wooden building provided plenty of space for design, construction, assembly, and testing, although, having been built for military training during World War I, it was on the austere side. At Dahlgren, a new wing was added to an existing building to provide space for Mark II and Mark III.

Mark II was built in sections so that it could easily be disassembled for shipment. Because the sections were too large to pass through the doorways in the Dahlgren building, however, provision had been made in the design of the new wing for a skylight large enough so that machine sections could be lowered through the opening by a crane.

Basic Speed of Operation

The internal speed of the machine was determined primarily by the operating times of the relays and the input tape reading time per instruction.

The operating time of a relay in either direction (pickup or dropout) was six to eleven milliseconds. The synchronous cycling controls (cam-operated contactors) allowed $16^{2}/_{3}$ milliseconds for relay operation, to provide a margin of safety. The time to carry out the most elementary operation was $2 \times 16^{2}/_{3} = 33^{1}/_{3}$ milliseconds. This was the time necessary to position the control relays and then transfer a numeric word through the bus, from one register to another. We will call this time, $^{1}/_{30}$ second, the Basic Transfer Time (BTT). The instruction tape readers delivered 30 instructions per second, or one instruction per BTT.

All the basic arithmetic operations and built-in functional calculations took many BTTs, even including addition/subtraction, because

Table 1

	btt	sec
Interregister transfer	1	0.033
Addition/subtraction	6	0.2
Multiplication	21	0.7
Division (x/y)	141	4.7
Reciprocal square root	180	6.0
Logarithm	156	5.2
Exponential	201	6.7
Cosine	225	7.5
Arctangent	285	9.5

of the use of the floating-point number representation, as described earlier. The basic operating times for Mark II are given in table 1 in terms of both BTTs and seconds. A comparison of the overall machine productivities of Mark I and Mark II will be given later.

Instruction Format and Programming

The instruction format for Mark II differed considerably from that used in Mark I. In general we tried to simplify and minimize as much as possible what the programmer had to write. Of course at this time higher-level languages were unknown, so that all programming had to be done directly in machine language. We established a rigidly formatted instruction cycle with a duration of one second in machine operating time, or 30 BTTs. During this one-second cycle time, the L or R machine could perform up to 6 interregister number word transfers, four floating-point additions/subtractions, and two floating-point multiplications, utilizing one transfer register, one adder/subtractor, and two multipliers, together with registers in the internal memory and the 46-wire transfer bus. (The ratios of 6 to 4 to 2 were chosen after counting frequencies of occurrence in sample problems.) Each interregister transfer required two instructions, and each add/subtract or multiply required three. At each of the 30 BTTs making up a one-second programming cycle, only one type of operation could be carried out: the programmer either called for that type of operation or skipped that particular BTT.

Of the 30 instructions (each corresponding to a BTT) that could be employed in the one-second programming cycle, 12 were allocated to beginning or completing an interregister transfer, 12 to reading an augend, addend, multiplicand, or multiplier from memory into the appropriate arithmetic unit, and 6 to reading the arithmetic result back to memory.

We can now estimate the relative speeds of Mark I and Mark II. Mark I took about 4.8 seconds to multiply two 16-digit fixed-point numbers, roughly the equivalent of ten-place floating-point numbers. An L or R machine in Mark II could perform up to two multiplications per second, but an average of 1.25 multiplications per second (0.8 second per multiplication) could reasonably be scheduled by the programmer in a typical problem. Normally no additional time was required in either machine for transfers and additions. Hence, the speed ratio was about 4.8 to 0.8—Mark II was about six times faster in internal operations.

A machine instruction defining the transfer of a numeric word was made up of four fields containing an algebraic sign code, two address codes (one for the source location and one for the destination), and an operation code. But in any instruction, one of the addresses was built into the programming format (and wired into the machine controls) so that the programmer didn't have to deal with it. Thus, the programmer had to define only one of the addresses: the other address was preestablished. The programmer wrote an instruction using six octal digits, which were represented in the coded instruction tape as three rows of holes, each row containing two binary-coded octal digits. The six octal digits were allocated as follows: one for algebraic sign, three for address (either source or destination), and two for operation. If the instruction represented a transfer from a storage register or input dial switch set to a functional register, the programmer wrote only the source address; if the instruction represented a transfer from a functional register to a storage register, the programmer wrote only the destination address. Hence, in the design we reduced by half the addresses that the programmer had to write, but we forced him to use a rigidly structured programming cycle.

The function of the algebraic sign field was to allow the programmer to transfer either X, the number as it was stored in the source location, or –X, Xabs., –Xabs., or X modified in accordance with the sign in a special "sign-invert" register. The address codes identified the 48 storage registers and 12 sets of input dial switches in each half (L or R) of the system. Most of the registers were used only for storage, but some also had special functions, being used to receive data from a punched tape reader, to deliver data to a tape punch or a page printer, to receive the argument to be used in a functional calculation, to perform certain control functions, or for other purposes.

The codes in the operation field were used to govern how input-output operations, built-in functional calculations, and other special operations were to be carried out.

The programmer was supplied programming forms or sheets. Each sheet represented one second of operation on the R or L machine. The sheet had 30 lines, numbered sequentially, corresponding to the 30 instructions that could be carried out in one programming cycle. Each line had space at the left so that the programmer could define algebraically the operation to be carried out, and spaces for a sign code, two addresses, and an operation code. In every line, however, either the source (out) address or the destination (in) address was already printed in using a descriptive word, namely: transfer, augend, addend, multiplicand 1 or 2, multiplier 1 or 2, sum or product 1 or 2.

Because of this highly structured form of programming, the number of pages in a written program represented also the number of seconds required to execute the program if branching operations or time-consuming input/output operations were not factors.

When the L and R machines were being operated separately and independently, the instruction tapes in a machine controlled only the complement of equipment in that machine. When the two machines were being operated as one composite calculator, any of the four instruction tapes could control the entire complement of equipment in the system. But another mode of composite operation was possible: here both an L instruction tape and an R instruction tape were running simultaneously, on different parts of the problem, each tape governing its machine. In the latter mode, there was a special provision for transferring numeric words between the L and R parts of the system. Four registers were provided in each machine to carry out conditional operations depending upon the algebraic sign of the number stored therein. These had the functions suggested by their names: "check," "sign invert," "start-stop," and "code interchange."

As mentioned earlier, the arithmetic operations performed directly by Mark II were only addition/subtraction and multiplication. To perform a division, take a square root, or compute any other algebraic or transcendental function the programmer needed to write a subroutine, but was given some assistance by the machine in the calculation of six functions, namely: reciprocal, reciprocal square root, logarithm, exponential, cosine, and arctangent. Assistance was also provided in searching and interpolating in tape-recorded tables of other functions.

For the two algebraic functions, an initial selection in a table of wired-in values produced a first approximation to the desired result. Then a programmed iterative procedure computed the final value. In the case of the four transcendental functions, initial calculations and selections in machine-stored functional tables were used to reduce the range of the argument to a small value. The reduced-range argument was fed into a power series calculation, and the result suitably combined with the previously selected values to provide the needed result.

Design Process

Having described in some detail the salient characteristics of Mark II, we need to discuss briefly the process used to design the machine and then to construct and test it. We begin with a question: Harvard University is not generally considered to be an organization that builds large equipment systems for other organizations to use, so how was it that Harvard carried out this activity? I think that three reasons can be mentioned. First, private industry was not yet involved in large-scale automatic computers, except for the work of BTL and IBM, so a university could provide a useful source of the necessary expertise. Second, the Mark II project was started under wartime conditions, by an organizational element of the Navy Bureau of Ships that was housed at, and supported by, Harvard. And third, Howard Aiken was, in addition to his other qualities, an effective entrepreneur and salesman.

Once the systems concept had been formulated, and the approach to component selection had been chosen, as discussed previously, the early phases of the design could proceed. These involved further definition of the system, final selection of the available standard components that could be utilized, and development of those components that had to be specially designed to meet Mark II requirements. The design of the family of relays that were used for storage, computation, and control has already been discussed.

In order to mechanize the peripheral (input/output) functions of the system, a combination of standard and custom-designed equipment were required. Equipment available from the Western Union Telegraph Company was eventually used, with only minor modification, for tape punches, low-speed tape readers, and page printers. Once the decision had been made to see whether "printing telegraph" type of equipment could be used, we contacted Western Union and had initial discussions with K. B. Mitchell, R. F. Dirks, and A. E. Frost of that organization. Although their equipment was rather slow, it seemed to meet other functional requirements, including adequate reliability, for many of our needs. Western Union was very cooperative, and Frost worked with us in defining exact specifications and acquiring sample equipment to test. Only minor modifications in selected standard equipment were necessary to fit in with our system design.

To supplement the Western Union equipment, we custom-designed high-speed handling and reading devices for WU tapes, devices that were used to introduce instructions and tables of functions into the system. We also custom-designed keyboards that could be used conveniently for manual entry of instructions and data words into the tapes. The keyboards allowed for the entry of a complete instruction or data word before the information was sent to the tape punch.

The high-speed instruction tape readers were designed to read 30 instructions, or 90 rows of six holes each, in one second. The reader was designed to sense mechanically 30 rows of holes three times per second, using a sensing "head" of $30 \times 6 = 180$ pins. Through the use of a geneva mechanism, the tape handling equipment was able to accelerate and decelerate the tape smoothly (so as not to tear the paper) three times per second, with pauses in between to allow the pins to sense the holes. An elaborate assembly of pulleys, tension arms, and tape position sensors was required to handle the tapes in an efficient manner.

A modification of the tape reading and handling equipment for instruction tapes was required for numeric function tapes. These needed to move rapidly in a forward direction for initial tape positioning, and then more slowly and intermittently in the reverse direction for readout from the tape.

The basic internal design of the system started with the circuitry associated with standard storage registers and means for reading out, reading in, and resetting to zero. Next the addition/subtraction and multiplication circuits were designed. These units required binary adders for exponents, coded decimal adders for the main ten-digit numbers, means for generating the five integral multiples of the multiplicand, and means for sensing, comparing, selecting, and shifting. Binary addition was performed in one pulse time, but coded decimal addition was designed to take two steps, in order to avoid having what we regarded as too many relay contacts in series.

A particular problem in floating-point addition was the roundoff required when digits to the right need to be dropped to stay within the ten-digit capacity of the system. We were able to eliminate both long carries upon roundoff, and bias, by doubling the rounding interval from 1/2 in the lowest-order place to 1 in that place. Whenever digits were dropped to the right, and the lowest-order digit retained was even, this digit was increased by 1. But when the lowest-order digit was odd, no correction was made.

A number of people participated in the design. The design of the tape mechanisms, together with all the structural and mechanical design, was carried out primarily by Robert E. Wilkins, assisted by Samuel T. Favor. I designed the basic addition and multiplication circuits. Others responsible for portions of the circuit design were Frederick G. Miller, Kenneth M. Lockerby, Charles H. Richards, Marshall Kincaid, and others. The circuit sketches made by these engineers were converted to finished circuit drawings and panel assembly diagrams by Lloyd C. Kentfield and Enoch Green.

Construction and Testing

In view of the size and complexity of the machine, and the large number of components involved, the construction and testing of the machine was a considerable undertaking, and required a large crew of people during most of 1946 and 1947. The team that had been developed to program for and operate Mark I supplied the core of people working initially on Mark II.

Frederick Miller was placed in overall charge of the final design stages and the construction and testing. Subsequently he went to Dahlgren to head up the testing and operation there.

William A. Porter, who had supervised the Navy operators of Mark I, supervised the Mark II construction team.

When in January 1947 I left the Computation Laboratory to join a newly forming computer engineering group at the Raytheon Manufacturing Company in Waltham, Massachusetts, my part of the design was completed, and had been formally documented by the electrical draftsmen. The fabrication and assembly work was well under way. In fact, various units of the machine were operated under test conditions during the Symposium on Large Scale Digital Calculating Machinery at Harvard, held that same month. Later that year, in August, the machine completed the solution of test problems. Its testing at Harvard having been completed, the machine was disassembled in late 1947 and packed for shipment. Mark II was received by Dahlgren in February 1948, and by September of that year it had been reassembled and checked out, and was solving ballistic problems.

Aiken's Alternative Number System

Henry Tropp

In September 1973, shortly before Howard Aiken died, I. Bernard Cohen and I interviewed him in his home in Fort Lauderdale. (For details concerning this interview and the difficulties we had in getting Aiken to agree to it, see *Portrait*. See also the chapter by Maurice Wilkes in the present volume. Tapes and a typed transcript are on deposit in the Harvard University Archives.) Professor Cohen left after the first two days, but I stayed on somewhat longer. Among the topics Aiken and I discussed were aspects of machine design, the choice of number systems as bases for computing machines, and the use of relays rather than vacuum tubes in the Mark I and Mark II machines.

One of the topics we discussed was Aiken's casual attitude toward the fact that others had taken out patents based on ideas for which he had clear priority of invention. When I raised the question of priorities, his response was colorfully expressed: "It is not true that if you build a better mousetrap, the world will beat a path to your door." Rather, to get people to adopt a useful innovation "you'll have to beat them over the head with a baseball bat."

When I bluntly asked Aiken to respond to those critics who said that the design of his last two machines (Mark III and Mark IV) were indicative of the fact that the world had already passed him by, he responded without offense or anger: "Hank, I was interested in getting working computers, not in having to invent new technology. Mark I used IBM technology because they built it. If RCA had built it, it would have been electronic." The originality of the design of Mark III, with its magnetic drums, contradict his statement, however. As the computer scientist Morris Rubinoff explained to me some years later, the magnetic drum memories were developed and built from the ground up. There had been nothing like them for Aiken and his collaborators to copy or develop.

One of the items that Aiken showed me was related to what he called an "electronic photographic printer." Aiken first expressed his belief that it had been his own invention, but he quickly corrected himself:

Aiken: Let's see. No, this was an invention of Harrison Fuller. This was one of the means of printing with cathode ray tubes.

Tropp: Now, was that intended for Mark II or had you planned to add it to Mark I, to replace the. . . .

Aiken: It was intended for Mark II and never used, and then reconsidered for Mark III, and also never used. These were circuits which generated the digits. By golly, multiple gate development! You see, people who say we didn't like electronics are in a little bit of trouble, because here it is, you see.

Tropp: Yes, and the date is 1946, I think.

Aiken: No. Let me see, what is the date on this? "On the Meriscope, 1947."

Tropp: January, 1947. So I was really only off by a month.

Aiken: Of course, this had to be going on for some months. So, here was an attempt to print electronically, and then there were multiple gates also in 1947. So you see, we were not so completely stupid.

Tropp: It's really curious, when you stop to think about the misconceptions that have been prevalent for two and a half decades now, about whose positions were what, when, and where.

Aiken: Our position [was] "By God, we had to have complete machines and they had to compute." And within that framework, we didn't give a damn whether we did it with carpet tacks or electronics or what. It didn't make any difference.

One evening I spent with the Aikens, I asked Howard if he could add anything to illuminate my understanding of how Mark I functioned. When I arrived the next morning, he suggested that I look at the open page of a yellow legal pad while he attended to something. On the page were lines of algebraic expressions. These were operation instructions, to be followed by looking up the value of a transcendental expression, interpolating, and then going on until there is a need to look up logarithmic values, again interpolating and inserting the value, before continuing. When Aiken returned to the study, I looked at him and said: "You are showing me that Mark I was designed to perform what you did on paper in a sequential, automatic process." His response: "Of course."

This led Aiken into a broad discourse on the advantages of using various number systems. (For more on Aiken's well-known opposition to the exclusive use of the binary system, see Wilkes's chapter below.) He handed me a book and a separate document. The book was a translation from Russian. He commented that it contained a discussion of a variety of number systems, and that the separate document was a proof that the optimal number system is one based on the transcendental number e:

Aiken: This is a discussion of all number systems from 2 to 12 and what they'd be good for, and the best representation of the radix-3 system. You don't use the digits 0, 1, 2, but you use 0, plus, and minus. Now you see, multiplication is extremely beautiful because it has no carries in it. And there are only two carries in this thing. [unclear phrase on tape] It's possible to record more than a thousand digits per inch.

[My response is unclear on the tape.]

Aiken: Yes, that's right, and this is a much more efficient system than the binary number system. I think if I were going to build a machine that was not decimal, I'd pick a trinary machine.

Aiken continued to stress the design advantage of base 3, suggesting that, on the basis of his analysis, a trinary design would "require less [word unclear on tape] than binary machines." All of this exchange was extemporaneous; neither of us was prepared to discuss this subject:

Aiken: The way you choose a coding system for your digits has enormous effect on the amount of apparatus that's used. Here is a base-3 scheme: It is where you want to multiply by 2 and the rating is 3. Let's take the digits 0, 1, 2 and now we want to multiply by 2. That goes 0, 2, 1. So you multiply by 2, and what you do is take that wire and that wire and interchange them. So if you get two wires coming down here like that representing the digits 0, 1, 2, in radix-3 system, you cross the wires like that and now you can multiply it by 2, and you tap off here, and that's the carry.

Tropp: I'll have to look at it again. It's so simple I can't believe it.

Aiken: Zero, 1, 2. Multiply by 2. Two times 0 is 0. Two times 1 is this representation, 1 over 2, and 2 times 2 is 1 to carry and 1 left over, which is odd 1. Now that column is exactly like these [phrase unclear]. So, here's the number x and here's the number $2x$ modulo 3, and there is only one carry and that's when there is a 1 here and so you tap off here and that's the carry.

Tropp: And that's the whole multiplication table.

Aiken: Yes. And the amount of apparatus is up to a place where you can get an add in the carry. It's zilch; it's zero. I mean in this representation of arithmetic with radix 3, you can multiply by 0, 1 or 2. And if you multiply by 1, you don't switch the wires; if you multiply by 2, you do switch them, and if you multiply by 0, you cut them off.

One aspect related to designs in other number bases that Aiken ignored was speed. He was willing to give up speed for clean, simple design and ease of use. He stated reasons for his preference for a design based on the decimal system unequivocally a few minutes later:

Aiken: Well, my choice would be to make it [the machine] decimal from the beginning to end. For commercial machines, for the machines that have to be commercially applied, or for individual use, I would use serial [design]. Throw away as much speed as possible to save the equipment and simplify it.

This was part of a discussion, only some of it recorded on tape. The following comment reveals an aspect of Aiken's thinking about computer design:

Aiken: Well, I think the place where I would change it if I could, I would express all programs algebraically, and I would extend what we ordinarily think of as algebra to include the manipulation of inequalities and signs of inequalities—lesser than, greater than—this sort of thing.

Tropp: To give you the proper branching and choice functions.

Aiken: And instead of regarding all of this as being something inherently associated with machines, I would regard it as part of the province of algebra, and I would teach people to think in terms of these signals exactly the way in which they think in terms of addition and subtraction and multiplication and division.

Tropp: And also in the way that they think of the truth tables for and-or disjunction.

Aiken: That's right, and with that all done, I think I could lay the ground-work to make programming [phrase unclear] rather than manipulation of [phrase unclear] that happen to be on a certain IBM punch.

In being willing to sacrifice speed, Aiken did not recognize the importance of speed in particular applications. He stuck to his guns on simplicity of design, precision of results, and machine reliability. He summarized some of the remarks he had just made as follows:

Aiken: One wonders if [we] couldn't have saved a good deal of magnetic tape if we'd gotten the trinary number system on tape, using a negative dipole for minus and a positive dipole for plus and no dipole for 0. In that way, we would have taken three digits to represent a decimal digit instead of 4, and similar extensions to take of letters [phrase unclear]. I believe that this search for speed was greatly overdone.

Tropp: Don't you think that that's still an overdone hangup that we have?

Aiken: Oh, yes.

Tropp: I mean, up to a certain point, you know it's like the supersonic aircraft; we never said to ourselves "Really, when we mean fast, how fast is fast enough?"

Aiken: Well, let's note that there are certain problems in defense and certain problems of nuclear physics in which the fastest computer we can build will never be fast enough.

Tropp: Okay, we'll grant those special cases.

Aiken: As long as we grant those special cases. Well now, as long as we are talking of commercial applications and the computational requirements of individual engineers and individual ventures, you don't need speed. . . . Now, to that end, the electronic computers that I built were always serial machines. You have the same equipment, and the equipment that you don't put into a machine never fails. So there was a lot to say for the serial machine: less equipment, not quite so fast, cheaper, simple. Of course, these little electronic hand boxes are all serial, so that's a throwback to the way we did it on Mark III. . . . Well, a million megacycles is still *pretty fast.* I've always thought that building binary computers for commercial applications was a mistake. You spend so much time translating from the decimal to binary number system and back that you would have been better off to do it in the decimal system. I argued that building binary computers is a concession to the designer to simplify his job—and the designer's job is done in a few months, but the user's job goes on forever after as long as the machine lasts, and really, it is his job that you are concerned with.

Another thing that came up in our discussions was the argument over whether there should be a fixed or a floating decimal point. I remarked to Aiken that this topic had once seemed very important. In the little compact electronic calculators, however, you now had a choice: "You can do your arithmetic either way." This gave rise to the following comment:

Aiken: Well, amusingly, I designed a machine that would add and, with end-around carry, subtract, multiply, and divide—and you needed only to push [a button]. You needed to respond to only one binary digit of information to tell the machine whether you wanted to operate in the binary number system or the decimal number system. I worked out the design of a machine to do that a year or so ago, more or less as an amusing exercise.

Aiken told me about a controversy—one of many—he had had with Norbert Wiener, who had said around 1947 that punched cards were obsolete:

Tropp: What was Wiener's position? Why was Wiener upset about the continued use of cards? Why did he feel they were obsolete?

Aiken: He was upset because magnetic tape had been invented and people were still using cards—and people who wouldn't give up on an old-fashioned procedure and go to new ones were stupid. So this was the basis of the argument.

Aiken then turned to some features of the computer's design that he believed to have been mistakes:

Well, cards are an excellent example of a mistake we made. We had cards with ten positions here—0 through 9—and we punched one hole, which represented a decimal digit. Now, if we had used four holes, we would have needed four-tenths the amount of paper, and by reading combinations of those four holes, we could have read decimal digits. I don't know how many billion cards have been produced operating on the one-out-of-ten coding system in the ten decimal digits. I suspect that the amount of paper wasted was sufficient to sacrifice an entire forest because of the stupidity. And we're still doing it! You still don't see any coded cards other than using one or two holes for a representation of the alphabet, which really is twelve holes in the cards. So it's with that exception—but we still represent the decimal digits by poking one hole out of ten.

II

Aiken's Program of Instruction and Training

Aiken and the Harvard "Comp Lab"

Frederick Brooks Jr.

In 1944, when I was 13 years old, I was reading *Time* in a public library. I came across an article announcing the dedication of a giant calculating machine at Harvard University. This was my introduction to Howard Aiken and his work, and indeed to the idea of a computing machine. Firing my imagination, it convinced me that I wanted to get into computing. As a undergraduate physics major at Duke University, I rounded out my studies of science and mathematics with more business-oriented classes in accounting and economics. From the *Time* article I had read years before, I retained the impression that Harvard was the center of opportunity for one who wanted to study computers, and that Aiken was the person to introduce a novice to the discipline. That is how I came to apply to Harvard for admission to the graduate program in computing established by the man I would come to call, affectionately, "the Boss."

I arrived at the Harvard "Comp Lab" in the summer of 1953 and immediately became acquainted with the Boss. Aiken, then 53, was at the height of his powers—alert, energetic, forceful, self-assured, and formidable. He was known to unleash his powers on waiters, airline clerks, and people in the laboratory. However, he liked spunk. Those of us who were refractory and difficult got along well with him.

Aiken was a marvelous teacher. He seemed to know his lectures by heart. He never looked at his notes, and his timing was impeccable. He would always finish the last syllable of a lecture, with a declining final inflection, two seconds before the bell rang.

Two settings dominate my memories of those years. The first is the office I shared with Ken Iverson while Aiken was my thesis advisor. When in town, Aiken would visit each of his advisees daily to read and discuss what we had recently written. We progressed rapidly in our work under this pressure, always wanting to have something new to show the Boss. Aiken was determined that we express ourselves clearly, thoughtfully, logically, and defensibly. His standard technique for guiding us was to throw out a batch of ideas to be explored and thought through. Some of them seemed to be lemons, and as soon as he left we would dismiss them; however, the next time he appeared, he would read over our new pages and ask "Haven't we lost an idea in here somewhere?"

It was a measure of Aiken's stature that he respected his students' results, conclusions, and writing. His willingness to respect students' conclusions is illustrated by an experience I had with my dissertation. Aiken had assigned me to investigate the design of machines specialized for certain business applications, an approach he was strongly convinced would yield substantial savings and improvements in performance. I was to focus on payroll accounting. After a year's work, it was very clear to me that machine specialized for payroll did not, in fact, have any advantage over machines designed for other serial tape file maintenance applications, such as utility billing or insurance premium accounting, but that these applications formed a general class for which specialized machines were indeed advantageous. Even though my conclusion went against his presuppositions, Aiken understood my view, respected it, and, I think, eventually adopted it.

Especially memorable from my days at the Comp Lab is the daily coffee hour. At 5 P.M. the crowd would gather in the machine room for coffee and wide-ranging discussion. That was where our business would be transacted. We had few occasions to enter Aiken's office. It was at a coffee hour that the Boss suggested that a payroll machine was the right topic for my thesis. It was there that he discussed the history of the computer laboratory and the development of the field of computing. It was there that he discussed the strategies and directions of the field. It was there that he discussed his philosophies about people, machines, and organization, and what distinguished him from his colleagues and competitors. It was there that he was a professor in the true sense of the word.

During these informal seminars, Aiken expressed his vision of computer architecture. Because of his emphasis on usability, he favored decimal machines. Because he stressed reliability, he preferred serial machines. He was committed to checked machines, and to conservative rather than "hot" engineering. He advocated a separate instruction store so there would be no way for a program to be corrupted by manufactured bugs. From an operational point of view this was sound practice—we manufactured our share of bugs.

One of my most vivid memories of the coffee hour is a discussion of patents. We asked Aiken why he had never challenged the Engineering Research Associates' patent for the magnetic drum, even though he was certain of his own priority in the invention of the device. "The problem in this business," he responded, "is not to keep people from stealing your ideas, but to make them steal them." He was never consumed with a passion for credit.

At one coffee hour, the history of the relationship with IBM and the relations between Aiken and Tom Watson Sr. came up. This was a source of deep bitterness on Aiken's part. I once had an opportunity to discuss this with Tom Watson Jr., and it was clear that this feeling was equally strong on both sides. The particular incident that triggered the greatest discord seems to have been the handling of the press at the time of the dedication of Mark I—especially the matter of the Harvard press release, issued with Aiken's collaboration before the official press conference. Watson perceived that Aiken had intentionally attempted to take sole credit for the development of Mark I; Aiken perceived the strife to have arisen from too great a degree of informality in handling the press.

There was, I think, a technical reason for the misunderstanding between Aiken and Watson. The distinctions among the architecture, the implementation, and the realization of a computer were by no means clear at the time. Though it is clear that what today we would call the architecture and a good deal of the implementation of Mark I had been designed by Aiken, the balance of the implementation and all of the realization had been designed by the IBM engineers Ben Durfee, Clair Lake, and Frank Hamilton. A distinction between those roles, had it been clear at the time, might have defused some of the bitter sentiment regarding the question of who deserved credit for Mark I.

By the mid 1950s, Aiken was completely on the outs with IBM. He was disappointed when he learned that I had decided to accept a job with the corporation. He let his disappointment be known, but he never held my choice against me. I was followed by Ken Iverson, Gerrit Blaauw, Peter Calingaert, and Bill Wright. Aiken remained cordial to us, despite our having "gone over." He was proud of us, and I suspect he had decided that the influence he had had on us might do IBM some good.

Perhaps Aiken's greatest legacy to his students was his staunch emphasis on computers as machines to be *used* and his consequent conservative approach to their design. This did not, however, prevent him from continually investigating new technologies, as is evident from his willingness to develop and use magnetic drums, magnetic tape, diode circuits, and magnetic-core shift registers. His architectural ideas were not so broadly influential as those of Eckert and Mauchly or those of von Neumann. Yet, in a sense, his machine architecture and implementation were way ahead of his time. For example, Aiken was so adamant about protecting proven program code that after he had recorded instructions on the drum of Mark IV he unplugged the write circuits so that there was no way to change the instructions. (In some respects, this was an early precursor to today's read-only memories.) Mixing data and instructions, according to Aiken, was a thoroughly bad idea of von Neumann's, and he was very outspoken in saying so. Today no one intermingles instructions and data.

That his ideas were not embraced to the degree that von Neumann's were did not bother Aiken. It was not important to him to belong to the "clubs" that characterized the discipline at the time. Indeed, in many ways it was important to him *not* to be a part of the paradigm that was evolving, but rather to maintain his own independence of thought. He felt that it gave him a lot of freedom to explore avenues that were not fashionable, an activity that would have been constrained had he been working in (or even with) one of the corporate laboratories. He relished the role of the independent outsider.

This attitude had its detrimental effects. I have often felt that Harvard sometimes had a "Charles River" view of the world and was not always willing even to look down the river. When I was at the Comp Lab, we were somewhat familiar with the work being done at MIT, but we were not at all familiar with the work going on at Princeton, Illinois, and Wisconsin, among other places. I believe strongly that the field would have advanced faster had we practiced the scientific tradition of scholarship that characterizes physics, a field in which colleagues openly share and disseminate their research.

Though Aiken's philosophy was not widely embraced in the realm of computer design, he was truly a pioneer in the field. For example, the degree to which today's computer science education programs mirror the program that Aiken already had in place by 1955 is amazing. The Harvard program in computer science included a two-semester course in numerical analysis and programming, a two-semester course in digital circuits and computer organization, a fair amount of advanced mathematics (not all of which computer scientists today would take, although it would do them no harm), a course in computer technology (taught by Bob Minnick when I was there), a course in analog computing (which one would elide today), a course in switching theory (a forerunner of digital logic courses) taught by Aiken, and a very valuable course in control systems engineering taught by Philippe Le Corbeiller (which I think Aiken encouraged us to take because he foresaw the importance of computers in control applications). My studies also included courses that would today form part of an applied mathematics or computer engineering program, such as courses in complex variables, probability and noise in circuits, boundary value problems, and mathematical linguistics. Aiken's program also included a course in business data processing introduced by Ken Iverson.

Many people think that Aiken was interested only in *scientific* computers. This was simply not so. During one coffee hour, Aiken turned to Ken Iverson, who had just finished his Ph.D., and said: "These machines are going to be immensely important for business, and I want you to prepare and teach a course in business data processing next fall." There had never been such a course anywhere in the world. Ken was qualified only because he was a mathematician. I was so excited by the prospect that I immediately volunteered to be Ken's graduate teaching assistant.

Aiken was one of the first to realize the important potential of computers for business and to recognize that business applications would dominate scientific applications. Indeed, much of his philosophy arose from this insight. He insisted that business applications would require the use of decimal arithmetic. To ensure that a mathematical approach was taken toward the application of computers in the business realm, Aiken forged ties with utility companies, insurance companies, and other business organizations. Tony Oettinger, then a student at the Comp Lab, was assigned to study how banks' operations could be computerized. Gerry Salton worked with representatives of the Boston Gas Company. The lab conducted pilot studies of automated billing with the Edison Electric Institute. Aiken, through his pioneering work in computer education and business data processing, helped to steer young scholars into careers in computing and to steer businesses into the information age.

In assessing Aiken's legacy, one must, of course, examine the influence of his students. The careers of his Ph.D. students illustrate the rippling waves of Aiken's influence. Herb Mitchell went to work for Univac in software. Gerrit Blaauw, after serving as one of the architects of the IBM System/360 family, became professor of digital technique at the Technical University of Twente, in the Netherlands. Charles Coolidge established his own computer company. Bob Minnick became a professor at Rice and later president of his own firm. Ken Iverson, who received the Turing Award for his development of APL, went on to IBM. Warren Seaman became dean of the School of Computer and Information Science at Syracuse, having been head of applied mathematics at Sperry and then at the Burroughs Research Center. Tony Oettinger became Aiken's heir at Harvard. Peter Calingaert had a fruitful career at the University of North Carolina. Leroy Martin became Assistant Provost for Computing and a professor of computer science at North Carolina State. Bob Ashenhurst had a productive career at the University of Chicago. Albert Hopkins went to the MIT Instrumentation Lab, which became the Charles Stark Draper Lab, where he eventually became assistant director and established a reputation in fault-tolerant computing, and later moved on to ITP Computing. Gerry Salton became a professor of computer science at Cornell and a preeminent figure in the field of document and information retrieval.[1]

Aiken was not only a goad and a guide while his students were at school; he continued to be so throughout their lives. For example, in 1970, in celebration of Aiken's seventieth birthday, a group from the Comp Lab congregated at the University of North Carolina, where I had invited Aiken to present two lectures. Though for five years I had managed to balance my research and teaching duties with my administrative duties, by 1970 the chairmanship of a growing department was becoming increasingly burdensome and was cutting into the time I had available for research and instruction. I confided my frustrations to Aiken. Always one to cut to the quick, he said: "Get yourself an associate department chairman. It doesn't have to be a computer scientist, just a person who is a skilled and able administrator." The latter point would never have occurred to me, but Aiken was right. I can recall many instances over the years after I left his tutelage in which Aiken's counsel, his quick grasp of problems, and his ready understanding of forces within institutions proved valuable to me.

1. Fred Brooks, after years at IBM, where he headed the development of the operating system for System/360, went on to found and head the Department of Computer Science at the University of North Carolina at Chapel Hill. —I.B.C.

Aiken as a Teacher

Peter Calingaert

Howard Aiken was a masterly and deliberate teacher. He designed and taught at multiple levels: curriculum, course preparation and delivery, examinations, research supervision, and mentoring. In teaching, he employed multiple tools: the environment of the Computation Laboratory, the formal lecture, the face-to face conference (or one-on-one confrontation), and the multi-person conference.

Curriculum

Like other early teachers of computer science, Aiken had to design a curriculum *ab initio*. There were no guidelines. The Association for Computing Machinery, established in 1947, would not promulgate its first curriculum until 1968.

The degrees for which Aiken and his colleagues recommended students were graduate degrees in applied mathematics, and hence they were subject to the policies of Harvard's Division of Applied Science.[1] The division imposed few academic constraints on its research groups, and it is probable that the choice of courses was essentially Aiken's.

Aiken's background in the electrical power-generation industry had given him a respect for practice. His background in pure and applied mathematics had given him a respect for theory, as had his success in applying theory to the design of computers. It is not surprising, therefore, that his curriculum incorporated substantial elements of both theory and practice.

---

1. The division has often been renamed and reorganized. Among the names that have been used are Department of Engineering Sciences and Applied Physics (in the postwar years), Division of Engineering and Applied Physics, and Division of Engineering and Applied Sciences. The last of these is the current name.

I do not recall seeing a written list of requirements. As the sole academic adviser until the late 1950s, "the Boss" didn't need one. Required courses taught at the Comp Lab were his own semester course in computer design and switching theory, entitled Organization of Large-Scale Calculating Machinery through 1954 and Switching Theory thereafter, and the two-semester sequence in Numerical Analysis. The two-semester sequence in Data Processing became required upon its introduction in 1954.

Other required courses were Applied Function Theory (taught by the classical applied mathematics faculty of the Division of Applied Science); Electric Circuit Analysis (taught by its applied physics faculty), and a pair of semester courses on digital electronics (taught by Harry Mimno of the Cruft Laboratory). In the first semester, Timing Circuits and Pulse Techniques covered digital circuitry in general; in the second, Electronic Control and Calculating Circuits took us through a Loran receiver, a radar transceiver, and a digital computer— one vacuum tube at a time. Frequently recommended were Philippe Le Corbeiller's course on Servomechanisms and Feedback Systems and a course in Applied Probability and Introduction to Random Processes, which had various teachers.

The Numerical Analysis course included weekly laboratory exercises using electromechanical desk calculators, as well as occasional projects using the Comp Lab's most modern mainframe computer. There was no course in programming as such; the technique and the art of programming were simply worked into the Numerical Analysis course.

Nor, despite the older title of Aiken's course, was there a course in computer organization *per se*. Although his course ignored the input/output subsystem and spent little time on problems of overall control, it did address issues of component choice and computer system structure. Moreover, we all learned about computer organization by informally discussing every new computer as soon as it was announced.

Today most of the subjects in Aiken's curriculum are routinely covered in undergraduate courses, but such was not the case in the 1950s. Undergraduates rarely appeared at the Comp Lab until 1956, when we introduced a course for third- and fourth-year students. Even then, undergraduates did not participate significantly in other Comp Lab activities.

Teaching Courses

Books on computer topics were rare in the 1950s and uncommon until the late 1960s. What we would now call computer textbooks were virtually nonexistent. Many of today's students, teachers, and practitioners can hardly imagine what it was like not to have a rich choice of books on almost any computer topic.

Research results were reported in the archival literature, principally in the *Transactions on Electronic Computers* of the Institute of Radio Engineers (later merged into the Institute of Electrical and Electronics Engineers). Although we were rarely assigned articles in the literature, we all managed to read each issue of the *Transactions*.

For his Switching Theory course, Aiken used as a textbook the research monograph *Synthesis of Electronic Computing and Control Circuits*, the nominal author of which was "The Staff of the Computation Laboratory." That designation didn't fool us; we knew who was the principal author. The book was volume XXVII of the Annals of the Computation Laboratory of Harvard University, published in 1951 by Harvard University Press. We called it, simply, "volume 27."

Aiken's written assignments were not unusual. Like many other assignments in technical courses, his problem sets included both "finger exercises" (intended to develop technique and fluency) and thought problems (intended to aid in the mastery of concepts). One respect in which volume 27 is a monograph rather than a textbook is that it does not include exercises. I do not know whether Aiken reused the same problems in successive offerings of the course. We certainly learned from those he assigned the semester I took it.

The crowning glory of a course with Aiken was the classroom experience. Even though most of the substance of the lecture material was in volume 27, we always looked forward to the lecture itself. Applied Mathematics 218 met each spring semester at 9 A.M. on Mondays, Wednesdays, and Fridays. Although Aiken was out of town an average of one business day in three, he scheduled most of his trips so he would be able to meet with the class in person (though occasionally his teaching assistant or another person would lecture).

Class began punctually at 9:07 (that's what "nine o'clock" was defined to mean at Harvard) and ended punctually at 10:00. Aiken wore a suit and a white shirt, as he did for virtually all appearances outside his office. Although there was occasionally some discussion, and questions from the class were accepted and answered with courtesy, class was by and large a formal lecture.

And was it smooth! Aiken lectured with minimal reference to notes, perhaps a few lines on a card. He claimed to have prepared each lecture during the 25-minute drive to the Comp Lab from his home in Winchester. It is true that he had been teaching Switching Theory for several years, and that he was himself the developer of the algebraic approach that underlay the course. The lectures were so good, however, that I cannot imagine that substantial preparation had not preceded the morning drive.

To claim that Aiken mesmerized us would be to overstate the case, but not by much. His was an imposing presence; one could sense his entry at the opposite end of a 60-person cocktail party. Standing in front of a class of perhaps two dozen students, Aiken dominated. His delivery was that of a retelling of a discovery. He made us feel that we were at the very frontier of research, even when covering material known a century earlier to George Boole.

Aiken's blackboard technique was smooth and unobtrusive—no multiple colors, no continuation lines in remote corners, no errors to be erased and corrected. He wrote what was helpful, and no more. He wrote legibly, and he left the material on the board long enough to be understood and copied. He spoke to the class, not to the board.

Aiken made excellent use of his voice, varying both pitch and intensity to maintain interest and offer contrast. He stressed the most important points by dropping the intensity to the point that he was barely audible.

The most remarkable feature, however, was Aiken's timing. As the clock approached 10:00, the intellectual excitement would rise. He would conclude with a climactic statement, the bell would ring, and he would slip out the door, leaving us eager for more.

Examinations

Like some other teachers, Aiken thought examinations were meant to instruct as well as to measure the student. This is easier to accomplish in oral than in written examinations, but Aiken tried in both.

Written Examinations
In addition to the written final examinations mandated by the university, instructors often gave one or more written course examinations

during the semester. Aiken, however, ordinarily did not do so. His finals were closed-book, and the questions were usually brief. Questions requiring essay answers, such as design sketches or comparisons of competing approaches, were common.

Four of Aiken's 3-hour final exams are reproduced here (minus the so-called minimizing charts and decomposition charts).

Of the examinations I have, only the one from June 1949 offers any choice of questions. That exam emphasizes computer system and subsystem design much more heavily than switching theory. Only question 3 addresses the design of a combinational switching circuit.

The June 1952 exam shows clearly the influence of volume 27. Half of the questions (1, 2, and 4) are purely on switching theory, and they are based on the book. Questions 3 and 5 also involve mathematical ideas.

Computer design as such is missing from the 1955 exam. Question 2 may be an example of using an examination question to instruct. Question 4 marks the first appearance in Aiken's examinations of a major extension to his own switching theory. The theory of functional decompositions of switching functions was developed by Bob Ashenhurst, who by 1956–1957 was teaching a sequel to Aiken's course.

The 1956 exam is the latest one without a significant admixture of questions contributed by others among us, especially Bob Ashenhurst and Warren Semon.

Oral Examinations

Oral examinations were conducted by committees of three professors. Aiken maintained close working relationships with few professors in the Division of Engineering and Applied Science. Two exceptions were Harry Mimno and Philippe Le Corbeiller, whom he often invited to serve with him on exam committees. Several of us faced that particular triumvirate.

The Ph.D. requirements at the time were to demonstrate mastery of four fields of knowledge, two by course work, one by a thesis, and one by oral examination. The only oral examinations were this qualifying examination and the thesis defense. The latter, with rare exceptions, was more often a presentation than a true examination.

Because the thesis was invariably in the student's area of research specialization, the qualifying oral examination was in a secondary area. That very fact added spice to what is traditionally an exciting moment in a graduate student's career. We normally felt somewhat intimidated

1948–49

HARVARD UNIVERSITY

APPLIED SCIENCE 218

1) Describe three methods for multiplication, and give single line diagrams for the corresponding multiplier circuits.

2) Compare serial and parallel computers and give the relative advantages and disadvantages of each.

3) Design a vacuum tube circuit of four inputs having outputs described as follows:

(a) The first output wire is to deliver a high voltage when the first and third input wires are low, and the second and fourth inputs are high. The first output wire is also high when the first and fourth inputs are low and the second input is high. The first output is to be a low voltage in all other cases.

(b) The second output is to be low whenever the first input is low and the second input is high. This output is to be high in all other cases.

4) Describe four of the following means of number storage, and give the relative advantages and disadvantages of each:

(a) Acoustic Delay Lines

(b) Selectrons

(c) Magnetic Drums

(d) Static Magnetic Delay Lines

(e) Cathode Ray Tubes

(f) Relays and Trigger Pairs

(g) Perforated Paper Tape

(h) Magnetic Coated Paper Tape

5) A magnetic drum computer stores 50 numbers per channel. Draw a single line diagram for transferring numbers from one channel to another with the aid of a transfer channel. Describe the use of any master or clock pulses required in the operation.

Figure 1
Examination paper, 1949.

6) Explain how pulses representing binary digits can be read into and out of a magnetic core, and show how this basic operation may be used to construct a magnetic delay line.

7) Discuss the general organization of Mark III calculator.

Choose any six questions

Final. June, 1949.

when facing the Boss on any occasion. Moreover, the American educational system has never offered practice in taking oral examinations. Anticipating one at the Comp Lab could be traumatic. Aiken, aware of this, worked to put the student at ease. He usually began with a question that he knew even the most tongue-tied student could answer: "How did you prepare for this examination?" Once (in Bill Kellogg's examination, if I recall correctly), he continued by recounting a visit as consultant to an industrial research laboratory, pointed out an observation that he had made, and asked what advice Bill would have offered the company. After Bill had coped with the challenge, Aiken couldn't help stating that the advice he actually offered was to discharge the director of research!

Aiken also often asked the student in what area under examination he felt best prepared. I hope Ramón Alonso will forgive me for recounting the following. In an oral examination on electricity and magnetism, Ramón was asked to choose a topic; he picked rotating electrical machinery. No one had told him of Aiken's earlier career in the power industry. Aiken asked for an analysis of the homopolar generator—indeed rotating, electrical, and machinery, but surely not the kind of generator found at your local power station. Ramón passed the examination, but we all kidded him about his choice.

Aiken could be fierce, even in an examination, if he felt that a student did not meet his standards. According to a widely told story, a student once said "I know where to look that up" in response to a question. Aiken was said to have replied "Go away and don't come back until you have."

My own qualifying oral illustrates Aiken's desire to teach during an examination (I left knowing much more about Ricatti's equation than I had known on arrival) and his grounding in practice. He asked me what advantage one integration formula had over another, despite a larger error term. The answer was that the symmetry of the coefficients permitted terms to be added in pairs before multiplication by the coefficients, thus halving the number of multiplications.

Research Supervision

The M.S. degree in the Division of Engineering and Applied Science required no thesis and only eight semester courses, five of which had to be at the graduate-only level. A well-prepared full-time student could earn the degree in one academic year. Even someone with a

1951–52

HARVARD UNIVERSITY

APPLIED SCIENCE 218

Professor H. H. AIKEN

1. Discuss the operators T_n, C_n, P_2, F_n, and B_n, and show that they represent the logical behavior of multiple triodes, multiple cathode followers, pentodes, forward rectifiers, and backward rectifiers, respectively.

2. Discuss the theoretical basis for the construction of a three variable minimizing chart.

3. Show that simple circuits can be constructed to form the quantities $2x$, $\frac{1}{2}x$, $5x$, and $\frac{1}{5}x$, when x is represented in a coded decimal system.

4. Find the minimal vacuum-tube operators for the functions

$$f(w,x,y,z) = w'x'y'z' + w'x'y'z + w'xyz' + wxyz'$$
$$f(w,x,y,z) = w'x'y'z' + w'x'yz + w'xy'z + w'xyz'.$$

5. Describe one decimal column of a Mark I storage counter from electrical, mechanical, and mathematical points of view.

6. Describe the shift and normalizing circuit of the Mark IV Calculator. Include a single-line block diagram.

Final. June, 1952.

Figure 2
Examination paper, 1952.

1954-55

HARVARD UNIVERSITY

APPLIED MATHEMATICS 218

1.   (a)   A doubling circuit is required in a machine operating exclusively in the 5421 coded-decimal system. Derive
algebraic or tabular expressions for the binary digits of a
single decimal digit of the product, as functions of the
multiplicand digits.

(b)   The same for multiplication by 5.

(c)   Can similar circuits be obtained for multiplication
by 3? Explain.

2.   The addition of ternary digits may be defined according
to the following table:

addend digit	augend digit	carry digit	sum digit
0	0	0	0
0	1	0	1
0	2	0	2
1	0	0	1
1	1	0	2
1	2	1	0
2	0	0	2
2	1	1	0
2	2	1	1

Figure 3
Examination paper, 1955.

Let the ternary digits be represented by binary digits as follows:

0	00
1	01
2	10

Label the binary digits of the addend w and x, those of the augend y and z, letting w and y be the high order digits.

(a) Obtain canonical forms for functions $f(w,x,y,z)$ and $g(w,x,y,z)$ representing the high order and the low order binary digits of the ternary sum digit, respectively.

(b) Use minimizing charts to obtain minimal vacuum tube operators for the functions f and g.

3. (a) Sketch a block diagram for a decimal adder with end-around carry, clearly indicating all carry paths.

(b) Two n-digit binary numbers are to be added by means of a relay adder with end-around carry. Sketch the circuits required in detail sufficient to characterize the operation of the whole adder.

4. Discuss the theory of functional decomposition.

Final.   June, 1955.

1955-56

HARVARD UNIVERSITY

APPLIED MATHEMATICS 218

Spend no more than the indicated time on any one question until all have been attempted.

1. (30 *minutes*) Present four different binary coding systems for the representation of decimal digits and discuss their relative merits.

2. (40 *minutes*) Derive algebraic expressions for the sum and carry digits produced by a serial binary adder to which addend, augend, and carry digits are applied as inputs. Draw a minimal relay circuit for the adder.

3. (40 *minutes*) Discuss four methods of decimal multiplication and draw functional diagrams.

4. (40 *minutes*) Test for decomposition each of the following functions of four variables and derive for each a minimal vacuum tube operator. Sketch the corresponding circuits.

   (a) $f_i = 1$ for $i = 0,3,5,7,9,10,12,14$;

   (b) $f_i = 1$ for $i = 0,1,2,7,12,13$.

5. (30 *minutes*) Discuss the reading period assignment.

Final. May 1956.

Figure 4
Examination paper, 1956.

full-time job could study part-time and finish the degree in 4 years or less.

There were always a number of master's students at the Comp Lab. Aiken tolerated them, but had no real interest in master's students. To him, a real student was a researcher and a candidate for the doctorate. Until the late 1950s, Aiken was the only Ph.D. supervisor available at the Comp Lab. If you couldn't persuade him to take you on, you were out of luck. Moreover, if you started with him and he perceived (even incorrectly) that you had failed him in some way, you simply ceased to exist. Once on his blacklist, you were never off.

For those who avoided the pitfalls, however, it was a marvelous experience. Sixteen of us completed the Ph.D. under Aiken's supervision in the period 1948–1958.

The choices of thesis topics were conditioned by a very important problem facing early university teachers of computer science: legitimacy. It was essential for their students to have topics that were obviously respectable. This led in some instances to the choice of practical problems of computer or circuit design (which had the respectability of engineering) and in other instances to the choice of mathematical problems (which were deemed respectable whether practical or not). Software theses would not become respectable until much later. Aiken considered himself a practical man, and his students worked on topics that they thought at the time to be quite practical.

Aiken provided an environment of support and total immersion. He subscribed to the pressure-cooker theory of education: put students together in a pot, close the lid, and turn up the heat. Every research student had a desk, usually in a room shared with several other students. A few who were full-time Comp Lab staff members had private offices.

We were expected to be at our desks most of the time. Because we shared rooms, we talked much to one another. We probably learned as much from one another as we learned in the classes and laboratories.

In many ways, Aiken treated us as apprentices rather than as pupils. He made each of us feel like a member of the Comp Lab staff. We took part in the coffee-hour discussions. We were brought into meetings with all scientific visitors, including the most distinguished. I remember, in particular, extensive interchanges with Morris Rubinoff of the University of Pennsylvania's Moore School and with Anton van Wijngaarden of the Mathematisch Centrum in Amsterdam.

Financial support was provided for those who needed it, in the form of teaching assistantships during the academic year and research assistantships during the summer. In fact, Aiken expected us in summer either to stay at the Comp Lab and continue our research or to acquire industrial experience.

Research was supported chiefly through contracts rather than through grants. The US Air Force supported the operation of Mark I and Mark IV and the concomitant work in numerical analysis. Bell Telephone Laboratories supported the research in switching theory. Later, the American Gas Association and the Edison Electric Institute jointly supported studies in business data processing. All the contracts called for written quarterly progress reports.

For those of us in switching theory and computer design, a major event was the semiannual research meeting with Bell Labs. Once a year, the Bell Labs crew came to Cambridge. Once a year we went to New Jersey. These meetings were both technical and social. Junior and senior scholars spent two days in close proximity, getting to know one another and discussing the most current results and conjectures. Virtually everyone who attended was expected to make a formal presentation on his current research. Those of us from the Comp Lab often made presentations related to the sections we were writing for the forthcoming quarterly report. There was always lively discussion after a presentation, before attention shifted to the next. I rarely heard any discussion of budgets, which I presume were considered *in camera*. The meetings at Bell Labs often included brief visits to the research labs to observe systems under development.

A cocktail party and a dinner on the evening of the first day provided relaxation. The party would be at someone's residence. In greater Cambridge, it would be at Aiken's. In New Jersey, it would usually be at that of W. Deming Lewis, who for most of the period under discussion here was the manager responsible for the research contract. The dinner would be at a restaurant.

At one cocktail party in the autumn, the Boss announced to all within earshot—and that left out hardly anyone—that I was going to finish my degree the next spring. I'd had no intention or hope of doing so, but how could I fail Aiken after such a public announcement? He was a master of subtle pressure.

When Aiken was in Cambridge, he would roam the building. Graduate students had priority, and he came by our desks shortly after 8 A.M. Eschewing the morning (as most of today's graduate students in computer science do, perhaps because in the past it was easier to find empty machine cycles late at night) would not have worked in Aiken's laboratory: if one wasn't seen in the morning, one might as well not exist. Aiken would inevitably ask what progress we had made since his last visit, and he preferred to see something in writing. A self-proclaimed "reasonable man," he knew better than to expect real progress over each 24-hour interval. But if you went *two* days without reportable progress, you were suspect.

In the matter of professional credit, Aiken was extremely generous to us apprentices. Even if he had proposed the problem underlying an article or a technical report, and had contributed to its solution, his name did not appear. In volume 27, his name appears only at the head of the staff list and as a "signature" after the preface. One modest sentence on the last page of the preface is his only statement of claim to the major intellectual contribution of the monograph.

I cannot discuss the role of teaching assistants from personal experience. My impression is that Aiken gave them lots of responsibility, much freedom of action, and little supervision. He counted on them to learn by doing. Once he related this story: As a teaching assistant himself, he was wading through an extremely long laboratory report and found the notation "Aiken, are you reading?" He told us that he simply placed a check mark beside it and went on. This anecdote probably was part of his training of teaching assistants.

Postdoctoral Appointments

Upon finishing our Ph.Ds under Aiken, just over half of us stayed at the Comp Lab. A few who were full-time staff members continued to serve in their former capacities. All were given faculty appointments at the rank of instructor (on the tenure track) or lecturer (not on the tenure track). Giving assistant professorships to newly minted Ph.Ds was not yet the practice at Harvard, although it was reported to be current at a certain institution downriver. Several of us were indeed promoted to Assistant Professor, and Tony Oettinger was promoted to Associate Professor (with tenure).

Aiken's relations with the rest of the university were far from cordial, and it is remarkable that he was able to secure even the few initial faculty positions he did. In at least one instance he had to wait a year and, in the interim, make the appointee a contract-supported Research Fellow.

Using a Navy metaphor, Aiken would sometimes say "A good skipper takes care of his crew." For us junior postdocs, even those with the exalted title of Assistant Professor, Aiken's care included shielding us from Harvard's administration. The shielding was perhaps too effective; I, for one, learned very little of how Harvard really operated. Yet I have no doubt that Aiken had our interests at heart.

Harvard has a strong tradition of faculty independence. I once asked a senior colleague in the Division of Applied Science for specific advice on teaching a course. He was obviously reluctant to offer any, and visibly uncomfortable at even being invited to tell another Harvard professor, however junior, what or how to teach. Giving postdocs lots of rope was not only the pervading philosophy at Harvard; it was also Aiken's personal style. We had lots of communication with him, but we made our own academic and research decisions.

Three Unusual Facets of the Comp Lab

The environment Aiken established in the Comp Lab was further unusual in three unrelated respects. One was isolation from the bulk of research in progress at other institutions. A second was tremendous emphasis on clarity of written expression. The third was the Coffee Hour.

Research Isolation

Aiken distanced us from research in progress elsewhere by investigating it himself rather than enabling us to investigate. He rarely suggested that we make any professional visits. He discouraged membership in professional societies and attendance at conferences. On the other hand, he encouraged us to read technical reports and archival journals. He also arranged formal exchanges of reports with a selected few other laboratories.

Far from uninterested in what other laboratories were doing, Aiken spent much of his time out of town visiting them, often as a consultant. His object was not to hide other work from us, but to distill it and report selectively. It is difficult, of course, to judge to what extent he distilled, to what extent he filtered, and to what extent he blocked.

What was Aiken's rationale? I do not believe that he merely wanted to enhance his own position by being the purveyor of knowledge. Perhaps he felt that he could acquire knowledge more efficiently than we could. Perhaps he wanted to encourage us to think independently.

It might be supposed that our isolation would have made it difficult, later, for us to use machines other than Mark I and Mark IV. Yet Fred Brooks made substantial use of an IBM 650 during his thesis research in 1955 and 1956, and many of us learned readily to use the Univac that was donated to Harvard in 1956. After all, we had been studying not only the designs of other computers but also the techniques used in programming them.

Did Aiken's policy keep us apart from the mainstream of thought and development in the field? To some extent it did, but this was not always to our disadvantage. If there was a mainstream, it was not monolithic and certainly not always correct. The prevailing wisdom of the 1950s was to embrace John von Neumann's concept of a stored program whose instructions were modified arithmetically by other instructions. The separate storage of instructions and data in Aiken's machines made this gruesome practice impossible, and the substitutive index registers of Mark IV provided an elegant alternative. Today's prevailing wisdom praises the separate storage of instructions and data as the "Harvard architecture."

On the other hand, we were disadvantaged by our isolation from important developments in system programming and application programming. Much of the exciting work in what was then called "automatic programming" escaped our notice entirely. We lagged in our awareness of compilers, file maintenance, and sorting. On balance, our isolation was unfortunate and unnecessary, but it probably did no lasting harm.

Writing

Aiken insisted on crystal-clear writing. Himself a master, he taught mastery through successive reviews of our writings by Aiken and by our peers. Nothing was published in the name of the Comp Lab that had not been thoroughly vetted. There were no appointed editors, and there was no mechanism of formal approval. It was simply understood by all that careful reviews were essential.

Because of the quarterly progress reports, we had ample opportunity to sharpen our writing skills. My own first research report serves as an example. As a graduate of Exeter and Swarthmore, I thought I knew how to write. After my report was typed, I gave it to Warren Semon (then Associate Director of the Comp Lab). A day or so later, Warren returned it so covered with comments that I could hardly find the original text. I was furious, but later, as I read through his comments, I grudgingly observed that he had some valid points and that he had suggested a host of ways to improve the presentation. I rewrote the report, answering all Warren's objections and making some modest improvements of my own. The next reader was Ted Singer, whom I expected to make a few notes. The report came back as densely covered with comments as when Warren had checked it. I was at least as furious as before. Careful study convinced me, however, that Ted's points were as valid as Warren's. I prepared the third version with even greater care than the second, and took it to the Boss. Scanning it rapidly through his pince-nez glasses held as a lorgnette, he stopped at an offending word halfway down the first page. He explained his objection to the word, handed me the draft, and dismissed me. It turned out that each of my predecessors had had a similar experience with Aiken: rejection of a composition because of a single objectionable word. But I didn't learn that until after having written version four.

Aiken's insistence on clarity of exposition was as strong as his insistence on correctness of content. The latter was certainly true of his computations. Even his first computer, Mark I, had directly coupled printers to eliminate transcription errors. To my knowledge, no numerical error has ever been found in any of the nearly thirty volumes of tables produced by the Comp Lab.

Coffee Hour

At 5 P.M., when the working day was officially over, the professional staff would stand around a coffee pot in the machine room and talk. Although the Coffee Hour was perhaps more interesting when Aiken was in town than when he was out, it was held every day.

Everyone attended, from first-year students to senior staff. It was simply too important and too interesting to miss. The discussion was quite informal and far ranging. Topics might include a description of a project that one of us proposed to undertake, Aiken's distillation of the work at some laboratory he had just visited, or the details of a newly announced computer. (In those days we were familiar with all computers.) Someone might mention a specific research problem that he was wrestling with, and we would engage in what was much later glorified as brainstorming.

We joked sometimes that the Coffee Hour was designed to get an extra half hour's work out of us, but it was important for far more than that. It served as a clearing house for technical information, and as a vehicle for testing our powers of oral exposition. Most important, however, it helped to maintain camaraderie and esprit de corps.

Conclusion

Aiken skillfully integrated research into teaching. In fact, the two were nearly indissoluble. The establishment and maintenance of a research environment for us students was a major component of his educational philosophy. Research results appeared often in his classroom lectures. A notable example is his algebraic theory for the design of digital logic circuits based on vacuum tubes. He taught on a number of levels. Whether shaping a curriculum, preparing a lecture, or conducting an oral examination, he was teaching. Precepts and examples abounded. Nothing happened by accident. It may have appeared so, but I am firmly convinced that Aiken planned to a much greater extent than we realized at the time. Howard Aiken was a deliberate and masterly teacher.

Aiken's Program in a Harvard Setting

Gregory Welch (with a concluding note by Adam Rabb Cohen)

This chapter is drawn largely from Gregory Welch's Senior Honors Thesis, Computer Scientist Howard Hathaway Aiken: Reactionary or Visionary? *(Department of the History of Science, Harvard College, 1986). Most of the letters and documents quoted or cited are in the Aiken files in the Harvard University Archives, classified under two major heads: Aiken's correspondence (cited as "HUA, Correspondence") and the records of the Computation Laboratory ("HUA, Comp Lab").*

Adam Rabb Cohen, in To Discipline Computer Science *(Senior Honors Thesis, Department of the History of Science, Harvard College, 1990), compares the historical development of computer science programs at Harvard, the Massachusetts Institute of Technology, and Stanford University and identifies many of the reasons why, despite its early lead in the discipline, Harvard's program languished while MIT's and Stanford's programs flourished. Cohen's analysis bolsters the conclusion that Aiken planted his vision of a computer science program in soil in which it could not take root, in an institution that refused it the one essential nutrient it required: funding. An edited portion of his thesis is offered as a concluding note to this chapter.*

The Harvard program in computer science reached the zenith of its influence and prestige during Aiken's stewardship. Toward the end of Aiken's Harvard career, however, the program was somewhat in decline. The reasons for this are varied, but they essentially reduce to this: Aiken had planted the seed for a nascent computer science in soil most infertile. Harvard's complex and somewhat byzantine administrative system and tradition of detached scholarship were not congenial to the growth of the discipline within the university and discouraged seeking outside support for its expansion. When Aiken's at times acerbic personality, his tense relations with colleagues and with members of the Harvard administration, and his increasing divergence from the mainstream of his field are also considered, the failure of Aiken's vision to take root becomes easy to understand.

Planting the Seed of a New Science

In early 1946, when Aiken was discharged from his naval duties and returned to Harvard with the rank of a tenured Professor of Applied Mathematics, the success and fame of the IBM ASCC (Harvard Mark I) had established him as a leading expert in the developing field of computing. The fruition of Aiken's schemes and the attendant publicity defused the earlier skepticism of his colleagues, but not for long.

The general attitude toward applied physics and engineering at Harvard had changed as a result of the war. As applied sciences, however, these disciplines had a somewhat questionable position within the Harvard academic framework, with its ideals of disinterested scholarship and pure science. Into this unsettled environment Aiken strove to introduce an entirely new field of applied science. About 20 years later, this cross-disciplinary field would come to be known as computer science.

Through force of character, zeal, and drive, Aiken managed to establish a highly respected facility for computation, research, and instruction. He accomplished this without much direct support from the university by means of contracts with private and government institutions. Although Aiken constantly sought to increase the use of computers throughout the university, the Computation Laboratory remained a unique and fairly isolated operation. These factors, and the general difficulties experienced by practical programs within the university, placed the Computation Laboratory in an anomalous and precarious position.

Under Aiken's leadership, the "Comp Lab" pioneered teaching and basic research in computer science. Because of his aggressive tenacity, the lab flourished despite its anomalous position within Harvard. Aiken's fifteen-year struggle with the traditional forces of the university seems to have taken its toll, however, and in 1961 he decided to retire early from Harvard and start a new career.

Without Aiken's strong-willed guidance and advocacy, the fate of the Comp Lab was uncertain. Instead of a legacy of a strong center of computer science, Aiken left to Harvard a laboratory and a center of teaching and research that was rapidly falling behind programs at MIT, at Stanford, and at other universities.

Soil Most Infertile

Aiken developed his program in computer science in an intellectual environment that had undergone dramatic changes as a result of World War II. The development of radar and the atomic bomb had greatly increased the prestige of science, particularly in its applied disciplines. The war had also established a precedent for heavy government support of scientific research. Harvard felt the effect of this trend because it had been the center for several important military projects during the war, including, in addition to the Comp Lab, the Underwater Sound Laboratory and the Radio Research Laboratory.

The adoption of the new style of scientific research and its integration into the university system was a turbulent process at Harvard. Soon after the war's end, the Faculty of Arts and Sciences established a new Department of Engineering Sciences and Applied Physics, known as ESAP, which included individuals who were members of both the Physics Department and a separate autonomous unit for Communication Engineering, known as the Cruft Group, of which Aiken was a member. This was the latest in a long line of reorganizations that reflect the troubled position of applied sciences at Harvard. Aiken was fully aware of the questionable position of applied fields of science at Harvard; in a letter to John Lord O'Brian dated 17 January 1945 he wrote of discussions concerning "the future of applied science and engineering within the University." The continual reorganizations that took place over the years indicate that this question was never satisfactorily resolved. Though it appeared that applied sciences would enjoy an expanded position in the university, within 10 years their importance had been significantly reduced.

The erratic history of the Graduate School of Engineering—which existed as a separate entity sporadically for many years, sometimes offering undergraduate courses and sometimes not—is indicative of the constant controversy and turmoil that surrounded the place of engineering and applied sciences in the Harvard curriculum. In 1949, the Graduate School of Engineering was incorporated into the Faculty of Arts and Sciences (FAS) by a vote creating the Division of Engineering Sciences, which comprised the Department of Engineering and the Department of Engineering Sciences and Applied Physics. In 1951, the Division of Engineering Sciences became the Division of Applied Sciences and the two independent departments were dissolved. Three years later the division was renamed "Division of Engineering and Applied Physics," but it later reverted to its former name (which it retains to this day).

Equally illustrative of the troubled career of applied science and engineering at Harvard is the controversy aroused by the Gordon McKay bequest. McKay (1821–1903), a Harvard graduate, became a millionaire as a result of acquiring a patent for a means of stitching shoe soles to uppers. After improving the design to deal automatically with toes and heels, he made his fortune by leasing machines. His two main bequests were a foundation for the education of negro boys and a trust fund to be used for "applied science" at Harvard. After World War II, Harvard became embroiled in litigation when lawyers for McKay's estate claimed that the university was misusing the McKay endowment. As a countermeasure, Harvard established a the Division of Applied Science (stressing the name used in the McKay bequest) and assembled a committee of outstanding scientists and engineers to define areas of "applied science" appropriate to the terms of the bequest.

In the years immediately after World War II, science was on the rise at Harvard. In 1946, the university announced a grand scheme to increase its scientific facilities and promote scientific research, particularly in the more useful or applied areas. In a press conference announcing the project, Harlow Shapley, a famed astronomer and a prominent member of the Harvard scientific community, alluded to an underlying reason for the increased interest in science across the country and a motive for Harvard's own initiatives. It was all "very sweet and nice" to study poetry and the arts, he said, but the threat posed by the Russians could not be ignored. Poetry would not help with this problem. Science, on the other hand, had the capacity to save the world, and so Harvard had drawn up a "practical plan to lead in the advance of science."[1]

The core of the plan included the construction of a large complex of buildings, a "Science City," which would incorporate facilities for all the natural sciences. The first building, four stories high and 500 feet long, was constructed on Oxford Street near the physics buildings and named the Gordon McKay Laboratories. Although funds for the complete project were apparently available, and although the plan contin-

1. Catherine Coyne, "Harvard Plans for Deeper Science Study; New Buildings Announced" (HUA, Correspondence).

ued to be discussed for several years, the ambitious project was never completed in the form originally envisioned.

The notes of the Committee on New Appointments of Harvard's newly formed Department of Engineering Sciences and Applied Physics illustrate how the faculty of the department searched for an independent identity within the university. Addressing the hiring of new faculty, the committee stressed the need to define the "nature and scope of the Department's scientific and technological interests and activities." The committee resolved that its area of activity was "applied physics," which it specified as bridging the gap between pure physics and practical engineering and striking a "balance between theoretical and experimental work." This focus was defined so that the efforts of the department would "supplement existing work in the University."[2] This was the academic designation into which Aiken most nearly fit, straddling the classic division between "pure" science and practical engineering. The blending of these two extremes during World War II, when theoretical scientists collaborated with engineers to produce new technologies (such as the atom bomb) that used major leaps in theoretical knowledge, heralded the emergence of technology-based science. The new "big" science relied heavily on large-scale apparatus and was therefore very costly. The major source of funding for research came from the government, as an outgrowth of the military work sponsored during the war.

Among academics, particularly at Harvard, there was a fear that government funding would impose requirements and restrictions that would violate the scholarly aims of unfettered scientific research. Thus, the Committee on New Appointments of the ESAP Department issued guidelines for how the department should treat government sponsorship. Four restrictions were placed on any project receiving federal funds:

1. The project must coincide with the academic interest of the department.
2. The project must also receive university support.
3. The project must use university facilities.
4. The project must employ university staff.[3]

---

2. "A Preliminary Report of the Committee on New Appointments" (HUA, Correspondence, folder titled Minutes of the Meeting of the Department of Engineering Sciences and Applied Physics).

3. Ibid.

Obviously this was an effort to ensure a degree of autonomy and intellectual integrity in any academic research conducted with government support.

Harvard was very concerned about the character of government research. One policy it enforced in this regard was the prohibition of any projects that required faculty members to have access to classified information.[4] This ruled out many military projects. Despite its suspicion of government funds, the university recognized the importance of financing scientific research if its reputation in the sciences was to be maintained.

The Husbandry of a Vision

It was in the unsettled environment of postwar Harvard that Aiken sowed his plans for a center for computer science. The expense of running and acquiring computers forced Aiken to seek support from private industry and from government agencies. This activity ran counter to the Harvard culture, and when combined with the administration's perception of Aiken as an autocrat it alienated the Computation Laboratory from the university as a whole. With only minimal financial and institutional support from Harvard, and with Aiken's reputation in the computer area diminishing, the Comp Lab's fate and Aiken's were sealed. In this regard it must be remembered that in his final years at the Comp Lab Aiken was approaching the age of 60, and that many of his views must have seemed archaic to the bright young men who were coming to dominate his field.

As Aiken's reputation declined, so did that of the Comp Lab. Government support was the basis of the lab's existence. Though IBM constructed and donated the ASCC/Mark I to Harvard, the Navy financed its operation throughout the war years and established the Comp Lab. After the war, the military (first the Navy, then notably the Air Force) and other federal agencies continued to be the lab's primary sources of support.

The experience of the Navy's Bureau of Ships in producing useful work with Mark I motivated the Bureau of Ordnance to draw up a contract with Harvard (dated 1 February 1945) for Aiken and his group to construct a large computer for the Naval Proving Grounds at Dahlgren, Virginia. Aiken called the new machine Mark II. With

4. Anita Corson, memorandum, 24 March 1959 (HUA, Correspondence).

the war still raging, time was of the essence. Rather than experiment with electronic circuits, Aiken decided to stick with electromechanical technology. The degree of innovation in the new machine was, however, tremendous; it was in fact two separate computers linked together in such a way that they could be operated independently or in tandem. Mark II, possibly the largest relay computer ever built, was delivered in March 1948, just 37 months after the contract for its construction was signed. The speed with which Mark II was designed and constructed by a comparatively small staff and its great reliability[5] testify to the sagacity of Aiken's decision to use a familiar and proven technology.

At the end of the war, the Bureau of Ships discontinued its use of Mark I, and the Computation Laboratory became a Harvard facility. But the Bureau of Ordnance expanded its contract with Harvard to include operation of Mark I until Mark II was completed.[6] The growing staff of the Comp Lab and the increasing need for lab space in the three buildings of the Physics Department required that the Comp Lab be relocated from the basement of the Physics Research Laboratory to more capacious facilities. To house Mark I and the staff of the lab, Harvard constructed a new building adjacent to the physics buildings. When its construction was announced, Harvard hailed the new laboratory as the cornerstone of its planned "Science City." This announcement seemed to recognize the symbolic place of the Computation Laboratory in the university. Aiken had written that "the future of the physical sciences rests in mathematical reasoning directed by experiment."[7] It seemed appropriate that the Computation Laboratory serve as the foundation of a vast scientific complex, since "the basic nature of numerical procedures cuts horizontally across the vertical departmental organization of the University." It is ironic that, rather than being the center of the envisioned "Science City," the

5. Aiken, in a letter to Harold Seaton dated 6 August 1947 (HUA, Correspondence), reported that Mark II was tested by installing it next to a 16-inch artillery cannon. The computer was operated as rounds were fired from the gun. Though its glass case was shattered and the relays actually moved, the machine did not malfunction.

6. Howard H. Aiken, "Memorandum on the Computation Laboratory," 19 October 1949 (HUA: HUF 300.149.3).

7. Aiken, "Proposed Automatic Calculating Machine" (reprinted in present volume).

Comp Lab became an isolated building on the outer fringe of the university's science buildings.

Although officially the Computation Laboratory building was constructed with unrestricted university funds, in fact it appears that part of the funds came from the original $100,000 that IBM donated when it presented Harvard with the ASCC.[8] In addition, the construction cost was partially defrayed by the $4000 per month that the Navy paid to use Mark I during the war.[9]

In its design, the $250,000 Comp Lab building[10] reflected both Aiken's utilitarian impulses and the specifications given the architect by members of the lab's staff. It contained offices, lecture halls and classrooms, a machine shop, and a 60-by-60-foot computer room. The lobby was designed with public relations in mind. Large two-story windows permitted sunlight to stream into a reception area containing a small exhibit on the history of calculating devices. A glass wall separated the lobby from the computer room, allowing visitors to watch the computers in operation.

Cultivating the First Generation of Computer Scientists

Aiken had been driven into the field of automatic computing by the desire to obtain results. He realized early on that the ability to obtain results from calculations was restricted more by the ability to apply computers to problems than by the performance of the computers themselves. This realization prompted him to advocate and develop a curriculum in computing that stressed the operation and design of computers and methods of numerical analysis. The Navy agreed to sponsor Aiken's proposed program of instruction through its Office of Naval Research.

First offered in the fall term of 1947 by Harvard's Department of Engineering Sciences and Applied Physics, the one-year program granted a Master of Science degree in applied mathematics with special reference to computing machinery. Under the terms of the agreement, the Office of Naval Research nominated fifteen candidates who

8. Theodore S. Ruggles to Aiken, 16 March 1949 (HUA, Correspondence, folder titled John Price Jones Co.).

9. Aiken to Dean Haertlein, 28 September 1953 (HUA, Correspondence).

10. "Navy Building Gets Approval," *Boston Post,* 6 April 1946 (HUA, Correspondence).

met Harvard's qualifications, and Harvard selected another fifteen students for the program.[11]

In 1948 the Air Force, which had signed a large computer research and construction contract with the laboratory, agreed to underwrite the program. The prospective participants ranged from college students to professional engineers. In 1949, Aiken recorded that 76 individuals had enrolled in the program over two years, and 14 Master's degrees and one Ph.D. had been awarded. By 1953, the curriculum had expanded to allow students to focus on particular areas, ranging from the strict application of numerical analysis to the design of electronic calculating circuitry. The program gave students a solid background in the application of mathematical techniques and of the technologies used for computation.

The program, which combined instruction in the use of computing equipment with an introduction to the fundamental aspects of designing such equipment, was broad in scope and offered practical instruction. On 7 April 1947, Aiken wrote to a friend, Walter C. Beckjord, that the Department offered courses "in preparation for a technical career, without the technological drudgery of an engineering school." It thus reflected Aiken's philosophy that theory was useless without application and vice versa.

During the postwar years, Aiken embarked on a one-man propaganda campaign to spread the use of computers. Again, it was not computers *per se* that motivated him; it was the prospect of increasing human knowledge through their use. Aiken traveled widely, delivering speeches on the benefits and shortcomings of computers in various fields. He talked to Harvard clubs, colleges, professional societies, companies, and conferences. His presentations were well polished and excellently delivered. Aiken used slides to augment his oral presentations, and he crafted each speech to address the audience's interests. Aiken's missions took him around the world. He was particularly respected in Continental Europe. Several governments granted him awards, and many foreign scholars came to study under him at the Computation Laboratory.

Always interested in finding new applications for his machines, Aiken initiated contacts with members of other departments at Harvard to demonstrate the relevance of computers and numerical analysis to their interests. These efforts led to many pioneering applications

11. Aiken to Talcott Parsons, 20 August 1947 (HUA, Correspondence).

of computers at Harvard. In addition to a long list of scientific applications run on Mark I (many of them instigated by Aiken's colleagues in the Department of ESAP), several projects pushed the use of computers into new areas. For example, Professor Wassily Leontief of the Department of Economics developed a program that simulated the effects of varying economic conditions on national economies. (For this first application of computers in the social sciences, Leontief was awarded the Nobel Prize in economics. For more on Aiken and Leontief, see *Portrait*.) Aiken also sponsored a project in which Mark IV was used to compare texts of the Bible.[12] As early as 1949, Aiken discussed the idea of using a computer in legal research.[13] It is clear that, from a very early date, Aiken had a broad vision of how computers could be used as tools in diverse fields. Academic endeavors were only a part of the pioneering applications investigated on the early computers at the Computation Laboratory. Aiken cultivated extensive contacts within the business community, and under contract the Comp Lab performed basic research in computer technology and the application of computers to business problems. For example, soon after the end of the war an experiment was conducted in the use of computers to issue insurance premium bills.[14] Pursued on behalf of the Prudential Insurance Company, that experiment demonstrated the fundamental difference between business data processing and scientific applications of computation. The Comp Lab later performed research for utility companies on similar problems in computer billing.

Another way in which Aiken helped to spread knowledge about computers and their application was through two symposia on "Large-Scale Digital Calculating Machinery" held at Harvard. The first of these, held on 7–10 January 1947 on the occasion of the inauguration of the Computation Laboratory facility, was the first large meeting of professionals involved in the design of computers. Its primary focus was on technological approaches to machine computation. The second symposium (13–16 September 1949) had a decidedly different focus, evidencing Aiken's growing concentration on the application of com-

12. "Harvard Mechanical Brain to Read Bible," *Telegram*, Worcester, Massachusetts, 18 December 1951.

13. John M. Maguire to Aiken, 24 March 1949 (HUA, Correspondence).

14. Edmund C. Berkeley to Aiken, 25 August, 1947 (HUA, Correspondence). Although Mark I had a comparatively large capacity for calculation, its facilities for the input and output of data were not well suited to the production of bills.

puters rather than their construction and design. In fact, the 1949 symposium marks a major transition in Aiken's career. The first session was devoted to the presentation of current computer projects, much as the 1947 symposium had been. Benjamin Moore of the Comp Lab presented a paper on the Mark III computer currently under construction at Harvard for the Navy Bureau of Ordnance. Mark III was Aiken's first electronic computer; a second was soon to follow. Both of these later computers incorporated new high-speed magnetic memory devices.

In the introduction to the 1949 symposium, Aiken announced that the planned Mark IV would be the last machine he or Harvard would build; instead, his and the laboratory's efforts would be focused on basic research in the application of machinery to computation. The work would stress basic component research, system organization, computer programming, and applications of numerical methods in new fields—and the education and training of students in these areas. Aiken felt that the fundamental technological design of computers had progressed to the point where industry could take over their construction. *Newsweek*'s somewhat sensationalized account stressed the importance of Aiken's announcement:

It was almost as if Henry Ford had in the early 1900s told his fellow automobile manufacturers that he would never make another car but would spend the rest of his life on traffic safety problems. . . . Aiken had keynoted an abrupt shift in the emphasis of computer science. . . . The liaison between human being and the machine must be greatly improved. . . . There will have to be a new race of scientists like Aiken—links between human society and the robot brain.[15]

The remainder of the symposium consisted of presentations on the use of computers in a wide variety of fields, including economics, physiology, and psychology. In a very real sense, Aiken had consciously removed himself from the mainstream of computer science, which at the time was primarily involved in the design of improved computers.

Aiken's emphasis on pioneering new applications and facilitating the use of computers led to new and expanded programs of study in computer science at Harvard. In late 1953 Aiken proposed a program in the "subject of computing machinery with reference to data processing and office operations."[16] The next fall, the program was offered under the joint auspices of the Harvard Computation Laboratory and

15. "Revolution in Robotland," *Newsweek*, 26 September 1949, p. 58.
16. Aiken to James C. Messer, 3 November 1953 (HUA, Correspondence).

the Harvard Business School. Aiken discussed the motives for the program in a letter to a colleague:

Our reason for starting this new program is the recent advances in computing machinery and computing circuitry which have suggested that there exist almost wholly new applications to the program of data processing. It is our experience that the engineers who have been engaged in recent research are totally unfamiliar with the data processing program. . . . We at Harvard believe a strong educational program designed to overlap these two fields [engineering and accounting and office management] is of paramount importance if we are to succeed in the application of control circuitry and computing machinery in the near future.[17]

This program was the first ever specifically to address data processing as a division of computer science.

Despite its many successes, the Computation Laboratory held a somewhat anomalous position within the structure of Harvard University. It was not directly affiliated with any one recognized department, its goals and activities were often viewed as running counter to the spirit of the university, and it was very expensive to operate. As soon as World War II was over, one of Aiken's primary activities was obtaining funding for the laboratory. Harvard was unwilling to finance the operation of the computers and the laboratory activities, the cost of which Aiken estimated to be $100,000 a year in 1949. The search for outside support raised questions about how appropriate to a university the character of the work done at the laboratory was. Once Mark II was completed, the Navy no longer financed the operation of Mark I. Aiken signed a contract to perform work for the Air Force and the Atomic Energy Commission. These contracts, occupying the major part of the computer's time, hindered integrating the use of the computer with other aspects of research throughout the university.

It was a primary concern of Aiken and his peers that sufficient time for scientific research be available on the Mark I machine. For instance, when a university administrator wrote to the Navy granting an extension of the Navy's contract for the operation of Mark I, he specifically required that it "be possible for Harvard to have sufficient use of the machine to work out certain problems that may develop throughout the University, and likewise on occasions to use the computing machine for teaching demonstrations."[18]

17. Aiken to C. C. Chase, 5 April 1954 (HUA, Correspondence).
18. William H. Claflin Jr. to Rear Admiral G. F. Hussey Jr., 3 October 1945 (HUA, Correspondence).

When the Atomic Energy Commission and Air Force took over the operation of Mark I, a faculty committee was established to approve the projects to be performed. The contract, though it gave priority to AEC projects, was written to provide aid to basic research in scientific topics. The committee, composed of Harvard officials and faculty, judged the research proposals on the basis of their "intrinsic scientific value."[19] (Apparently, another reason the committee was established was to monitor Aiken's activities.) Though a substantial amount of academic research was conducted on the machines, the load became so heavy that in the autumn of 1949 the committee voted to accept no more proposals.[20]

Early on, Aiken stressed that the work for the military had intrinsic scientific value. He wrote the following to a Harvard administrator: "These problems are of the highest scientific importance and represent major parts of the . . . program I would propose as a member of the faculty of the university even though all computational connections with the Navy Department were to be discontinued."

By 1949 it was apparent that things had to change if the Computation Laboratory was to become a truly independent academic center for research in computer science. Aiken prepared a long description of the history of the laboratory, including where it had gotten support in the past and what its needs were for the future. He saw the top priority to be the construction of a large computer for exclusive use by the university.[21] Mark I was obsolete, he pointed out, and would eventually no longer be needed by the federal agencies, which would be acquiring their own more modern facilities. Thus, federal funds could not be counted on indefinitely.

Since the Department of Engineering Science and Applied Physics supplied the laboratory with only $3000 of an estimated $100,000 annual budget, and since Aiken anticipated that it would take $400,000 to construct an upgraded computer and refurbish the laboratory, a new source of support had to be found. Aiken concluded that the university should seek private sponsorship. The past accomplishments

19. Edward Reynolds to W. E. Kelly, 10 December 1948 (HUA, Correspondence).

20. "Minutes of the Meeting of the Committee on Applied Mathematics and the Computation Laboratory," October 4, 1949" (HUA, Correspondence, folder titled Minutes of the Meeting of the Department ESAP 48–50).

21. Aiken, "Memorandum on the Computation Laboratory," 19 October 1949 (HUA: HUF 300.149.3), p. 8.

of the laboratory warranted such support, particularly since the laboratory would perform the "fundamentals of basic research which private industry can not do . . . [but] can, however, take the results of this basic research and make their own applications for the solving of their own problems."[22]

To investigate the feasibility of pursuing private subsidy of the laboratory, and to propose a strategy for doing so, Aiken and Harvard retained the John Price Jones Company of New York. That company conducted a thorough survey of companies that might be interested in supporting the laboratory, and Aiken and members of his staff visited numerous presidents and chairmen to sound out their receptivity to the idea. Aiken had numerous contacts in private industry, gained both through his reputation and through his consulting work, and the results of the survey testify to the respect he had earned in the private sector. However, the fund-raising project evidently came to naught. In 1950, Aiken wrote to John Price Jones: "We are still struggling along here at the Computation Laboratory with the aid of government funds. The day when we have private support is as far away as ever. Perhaps, though even this problem will be solved in time."[23]

Aiken managed to build his final computer, Mark IV, with development and construction support from the Air Force. Mark IV was in the Computation Laboratory, and Harvard had the option to pay for 35 percent of its operating costs and receive a proportionate amount of time on the machine.[24]

Aiken also found support for research at the laboratory through direct research contracts with private organizations. For instance, the Edison Electric Institute and the American Gas Institute each gave the laboratory $100,000 over three years in the mid 1950s. The contract with those two institutes was for "basic research to develop and evaluate the design theory of [computers to address] . . . the needs of . . . the electric and gas industries specifically."[25] The Comp Lab also per-

22. R. F. Duncan to Mr. Jones, 6 July 1949 (HUA, Correspondence).

23. Aiken to John Price Jones, 10 July 1950 (HUA, Correspondence).

24. Aiken to Paul Buck, 16 October 1951 (HUA, Correspondence).

25. H. S. Bennion and J. W. West Jr. to Edward Reynolds, 3 May 1955 (HUA, Correspondence). See also "Automatic Data Processing Progress Reports to the Electronic Research Steering Committee of the American Gas Association and Edison Electric Institute" (HUA, Comp Lab, HUF 300.806).

formed research for the Bell Telephone Laboratories, on a contract worth at most $10,000 per year.[26]

The difficulties of financing the operations of the Comp Lab and the means Aiken used to obtain support caused tension between Aiken and Harvard's administration. Aiken felt somewhat resentful about the lack of fiscal support from Harvard. In a note to his colleague Harlow Shapley, Aiken wrote of "the absence" of any "modern computing device at Harvard University at the present time, due in part to the penuriousness of the institution." Mark I was "still carrying on the major burden of the work."[27] Harvard was suspicious of the nature of the work at the laboratory and of Aiken's autocratic rule of the facility. When compiling materials with which to sell the idea of supporting the laboratory to private companies, the John Price Jones Company considered using the term "mass-production mathematics" to describe the work done there. However, in view of the laboratory's questionable position at the university, the company decided to discard the slogan on the ground that it had "certain unfortunate dangers with respect to the university's internal uncertainty whether the Computation Laboratory is conducting scientific or industrial enterprises."[28]

Aiken's personality and his management style also isolated the laboratory from the Harvard community at large. The close-knit environment of the laboratory contributed greatly to the productivity of Aiken's students, but it limited their contact with the rest of Harvard and with the outside world. Aiken was essentially the front man for the operation. He established the contacts and represented the laboratory to the outside world. This had an unfortunate impact on the reputation of the laboratory within the computer science community.

Diverging from the Mainstream

By the 1950s, Aiken began to be seen as a grating and conservative voice by the designers of state-of-the-art computers, and the insularity of the Computation Laboratory exacerbated this perception.[29] It was not simply his views that made Aiken unpopular; it was also the way he voiced them. One of Aiken's students recalled that, at a meeting of

26. M. J. Kelly to Edward Reynolds, 11 April 1952 (HUA, Correspondence).
27. Aiken to Harlow Shapley, 3 February 1950 (HUA, Correspondence).
28. John U. Monro to Aiken, 14 April 1949 (HUA, Correspondence).
29. See Maurice Wilkes's chapter in the present volume.

top computer experts discussing large computer construction projects Aiken stood up and facetiously advocated the advantages of "not" building computers, thus enabling the design be changed at any time. His audience was not amused.[30] Aiken merely wanted computer scientists to reflect on their discipline. His vociferous, reactionary, or humorous opinions were usually crafted for their shock value, to stimulate reflection—often to his detriment. "Aiken," one industry figure testified, "is not the best-liked person in the business . . . too blunt, too frank, too damn sure of himself, etc."[31]

Aiken had a very holistic vision of computers. Part of this resulted from his first involvement in the field. He saw himself as one of the last of the traditional builders of calculating equipment—scientists and mathematicians who built machines to further their work. As more and more computers were built, their designers no longer came from the ranks of users, and they had less touch with the actual end uses of the machines. They were specialists.

Aiken's attitudes were summed up in this comment: "The proving of machines is of real interest only to those to whom the advance of the state of the art is a major concern, while to the user for whom a computing system is but a research tool, such activity is at best a nuisance."[32] In the world of computer science, Aiken represented the user's concerns. He testified to a group at the Naval Research Laboratories in 1948 that, "although the machine operation speeds are comparatively small," they were "sufficiently fast until more is learned about numerical methods to be used in programming."[33] In a letter to the Bureau of Ordnance he elaborated on this philosophy:

The last three years have shown that reliability and dependability are the two most important attributes of a calculating machine, and these are followed by ease of operation. No amount of speed or other virtue will take the place of absolute confidence in the computed results. Computing speeds are already

30. Robert Ashenhurst at Pioneer Day. Source: taped record of Pioneer Day proceedings honoring Aiken and the Harvard Comp Lab at the 1988 AFIPS National Computer Conference in Anaheim. Copies have been deposited in the Harvard University Archives and at the Charles Babbage Institute.

31. Norman Hardy to Aiken, 16 December 1951 (HUA, Correspondence).

32. Robert Ernest Esch and Peter Calingaert, "UNIVAC Central Programming," October 1956 (HUA, Comp Lab, HUF 300.156.25).

33. "Synopsis of a talk given by Howard Aiken at the Naval Research Laboratory, Washington D.C., November 17, 1948" (HUA, Correspondence).

so fast that a gain of a few seconds in computing time is relatively unimportant in a general purpose calculator. On the other hand, the time required for problem preparation can be greatly reduced, and it is here that the greatest emphasis is being placed [by the laboratory].[34]

As these comments illustrate, Aiken's experience in running a computer center showed him that the primary bottleneck in the production of results was not the speed of the computer but the conversion of problems into calculable forms. At a time when immense effort and capital was still being poured into constructing faster computers, surely Aiken's opinions must have rubbed many the wrong way.

Since the Comp Lab lacked internal university support, its fortunes relied on its good relations with external sources of funding. As Aiken alienated himself from the mainstream of the academic and commercial computing communities, his personal situation jeopardized the health of the laboratory. In 1961, when Aiken retired from Harvard at the minimum age, the computer science program there had already begun to lag behind programs at other universities. Without an institutional commitment to nurture the program, computer science at Harvard dwindled rapidly. Aiken, like many other charismatic and forceful leaders throughout history, failed to establish institutions that could survive without his helmsmanship.

Concluding Note: The Decline of the Harvard Program

Adam Rabb Cohen

External funding has two sides. It can provide an academic program with the money and equipment to grow rapidly. At the same time, the program becomes dependent upon this support and hence vulnerable to its sponsors' mistakes and funding cuts. Harvard's computing program was more insulated from these external influences than many other similar programs, if only because Harvard's administrators discouraged its computer scientists from soliciting a large amount of external funding. They preferred that computer science be supported out of limited endowment money, which meant the program would always be poor and small relative to those that could "leverage" outside funds. This had structural consequences for the computing program,

---

34. Aiken, letter to the Chief of the Bureau of Ordnance, 24 November 1947 (HUA, Correspondence).

as scarce resources required it to remain a part of a larger Division of Engineering and Applied Physics. Computing could not secede to form an independent organization.

In contrast, computer scientists at Stanford, with an external support base of their own, were able to establish an independent department in the 1960s. This program grew quickly, surpassing the older one at Harvard in size and prestige by the 1970s. Like Stanford's, MIT's computing program won substantial support from government and industry; however, MIT did not create an autonomous department of computer science. At MIT, computer science and electrical engineering remained integrated in one enormous department (Electrical Engineering and Computer Science).

Over the long term, a program's size and structure affect its research and teaching capabilities and hence its reputation. Owing to the size of its program, Harvard has educated an order of magnitude fewer graduate students than Stanford or MIT, and its impact on the field has been corresponding small. The limited resources of Harvard's computing program have also encouraged it to concentrate on low-cost research at the theory end of computer science. The wealthier program at MIT, by contrast, has attempted to address the full range of topics.

Harvard has (with MIT) one of the two oldest academic traditions in computing, dating back to before World War II. Yet priority does not guarantee size or prestige. Stanford, a relative newcomer, has eclipsed Harvard in both areas. By the 1980s, Harvard had one of the smallest computer science doctoral program in the United States, while Stanford and MIT were giants in the field, each among the top 5 institutions in terms of the production of Ph.Ds in electrical engineering and computer science. And Stanford and MIT have reputations that match their size. In the 1980s, most surveys of graduate programs in computing gave them the highest ratings. In the same period, Harvard failed to place among the top 25.

It is often assumed] that scholars at an institution that relies on internal support for science always enjoy freedom to choose their research areas and their methods of proceeding, and that this freedom is risked when a university reaches out for support. This ignores that a university is composed of competing parties with conflicting interests. Within the internally funded university, it is those who control the internal funds—administrators, not scholars—who have freedom of action.

Harvard's computer science program has shown the effects of tight constraints on growth and total budget. These constraints stem from a desire among administrators to rely heavily on the existing endowment rather than on "soft" federal or corporate funding, as well as to keep computer science "in balance" with other academic efforts in the university. At Harvard the computing program's reduced reliance on outside grants insulated it from external influences but kept it small and made it more dependent on the administrators who controlled its internal funds. Because it was small and poorly funded, the program remained a loosely structured entity within the larger and wealthier Division of Engineering and Applied Physics (DEAP).

Keeping computer science within the DEAP, although it had financial advantage resulting from accounting practices peculiar to the Harvard bureaucracy and federal funding, had political drawbacks, however, some of which have been acknowledged by Harvey Brooks, Dean of the DEAP:

As long as computer science is within the Division, it tends to be regarded as of exclusive interest to the Division, and therefore not of much concern to the Faculty of Arts and Sciences as a whole. It ranks very low in the competition for the attention of the Dean of the Faculty and is not regarded as a possible candidate for faculty resources.

"Submerged" in the DEAP, computer science suffered from low visibility both inside and outside the university. Its ability to grow and attract new funds was limited, and its amorphous structure put off a number of tenure candidates who asked "Where's the department?" Indeed, the department struggled to survive and meet the growing demand by students for instruction in computers in a milieu in which the decision makers were largely indifferent to the fate of computer science at Harvard.

III

Recollections

Commander Aiken and My Favorite Computer

Grace Hopper

The following is an edited composite of a talk given at the Computer Museum and remarks made at the Pioneer Day ceremony honoring Aiken at the 1983 AFIPS National Computer Conference.

I met Howard Aiken and was introduced to a computer, the IBM ASCC/Harvard Mark I, in 1944, in the midst of World War II. I had been teaching mathematics at Vassar and had just been made a commissioned officer in the Navy and given my orders: I had the weekend to spend with my family, and then I was to report to the Bureau of Ships Computation Project at Harvard.

Finding the Computation Project proved no easy task. I eventually found an office where all the Navy operations and research at Harvard were coordinated; they had heard of the Project, but at first weren't sure where it was. Finally, at about 2 o'clock in the afternoon, I found the Computation Project in the basement of the Physics Research Laboratory!

When I walked in, Commander Aiken looked up at me and said "Where in the Hell have you been?" I tried to explain that I had been told I could take two days off and that I had just spent most of the day trying to find the place. "I mean for the last two months," he barked. "Midshipman's school" I answered. "Hell, I told them you didn't need that; we've got work to do." He waved his hand at Mark I, all 51 feet of her, and said "that's a computing engine."[1] After introducing me to Mark I, Aiken informed me that he would be delighted

1. The concept of a computing *engine* is something we seem to have forgotten. I think that when Aiken called it a "computing engine" he meant that it was made up of different parts, each performing a different function. And he was right, Mark I really was an engine: it had many parts which worked simultaneously, together with each other, and performed distinct functions. It took

to have the interpolation coefficients for the arctangent by Thursday. Two programmers were already on duty: Ensign Robert Campbell and Ensign Richard Bloch. They were most kind to me and helped me to get my first program onto the computer and to compute the arctangent coefficients as Aiken wanted.

Most people have forgotten why we were so greatly in need of computation during the war. World War II saw an almost complete change in our weapons systems. New weapons required new numerical techniques for their design and deployment, techniques that depended upon computation. For example, the Navy used new mines that either acoustically or magnetically detected the presence of a ship without direct contact. To deploy them effectively meant calculating the range over which the mines could damage a ship compared to the range at which they could detect it. The results of these calculations told us how far apart to sow the mines. A similar situation pertained with respect to depth charges. Early depth charges were merely rolled off the stern of a destroyer. The new ones introduced during the war were propelled by rockets. The question was where should they go and in what pattern should they be fired. Again, computations were required to find the answer. Self-propelled shells armed with proximity fuses posed similar questions.

All these new weapons required new computations, endless computations. Even after you found out where the mines would have to be positioned, you then had to know what kind and how strong a dipole was needed to detonate the mines later when you needed to clear them out of the way. Also, radar, loran, the A-bomb, and other new things that happened during World War II depended upon vast calculations. The pressure for that computation was very great; everything was "hurry up, do it yesterday." So the pressure to keep the only computer

---

in numbers, chewed them up like the dickens, and spewed out numbers again. It was more than a calculator—those lovely machines we had that sat on your desk where you entered numbers and ground them out by hand. Nor was it like the computers of today—black boxes, things on chips, one unit, one thing. But I'm convinced we will have to bring back the concept of a computing engine; as we start to build systems of computers, some of which will have specialized functions, we'll find ourselves going back to structures and to an architecture that is far more like Mark I's than we might ever have thought it would be.

we had running was very, very strong. Mark I ran 24 hours a day, seven days a week—a rough task for a small crew.

In my opinion, Mark I was the first large-scale, digital computer in the United States; the first machine to be built in order to assist the power of man's brain instead of the strength of his arm. Even though many more computers were built during the war and later, Mark I will always remain the first in the United States. We are apt to forget that Mark I was automatically sequenced, which is to say, she was programmed. Great praise is bestowed upon ENIAC as our first electronic computer, but ENIAC was not originally programmed; that capacity was only added later. ENIAC was a plugboard machine on which you simulated a problem. So, from the point of view of the programmers, because it was sequentially programmed, step by step, one operation after another, Mark I clearly resembled more closely what we have today than other early machines.

Similar as it was to modern machines, Mark I incorporated one feature that today's sequenced program computers do not have: if you did not explicitly tell Mark I to go ahead and execute the next instruction, it stopped. Programmers had to enter a 7 after each command to instruct the computer to proceed. If she didn't get that signal, or some other automatic signal, she just stopped dead and waited for you to tell her what to do. From some points of view this was a tremendous advantage; for example, it was wonderful for debugging. To the best of my knowledge Mark I was the only sequenced program computer to include this feature.

Mark I was an engine consisting of several parts or units. Like Babbage's Analytic Engine, Mark I had a "store" where you entered numbers, and a "mill," or (as it was called on Mark I) a sequence mechanism, that worked on the numbers. The sequence mechanism could also call upon other units, such as the multiply/divide unit to perform specific operations.

Mark I's instructions were essentially single-address instructions, though no one ever called them that at the time. We wrote all instruction in three columns: the first column (A) was the "out" column, the second (B) was the "in" column, and the C column was the "action" column. The "in" column was essentially the operation to be performed, the "out" column was the location of the number to be affected, and the "action" column merely told the machine what to do next. So it really was a single address instruction, where you named the quantity first and then the operation. There were various action codes that could be used in the third column along with the automatic advance code 7. The third column was the one that, for instance, would tell the computer to print or punch a card, or perform other such operations.

In Mark I the data and the instructions were totally independent. Throughout the development of Marks I, II, III, and IV, Aiken always insisted that the data and the program must be stored independently. We lost that concept for a while when people came along and said "Oh, we want to store the program in the same memory as the numbers, so that we can alter the program." In my opinion, that put more bugs in programs than anything else ever did. Only in the last couple of years have we come back to recognize the importance of keeping the data and the program independent and in separate memories, one acting on the other. Perhaps in the future more computers will come to reflect this feature of Mark I and of Aiken's philosophy.

Mark I's structure forced us to adopt some unusual mathematics. For example, the machine distinguished between positive zero and negative zero. Plus zero was all zeros, minus zero was all nines. This was a by-product of the fact that we used nines-complement arithmetic to execute subtraction since the machine contained adders, but no "subtractors" as such. Despite the insistence by one of the staff mathematicians clear up until the end of the war that was no such thing as plus or minus zero, only zero, Mark I and I both recognized the difference. Mark I's check register tested on the sign of a quantity. So, every so often when a program had crashed, Aiken would tell me "Go fix that." The first thing I always checked was if the program had tested on zero and gotten the wrong one. It usually had. To me it made perfect sense. If you came down in the positive quadrant, the positive numbers get smaller and smaller as they approach zero. Eventually they drip off the end of the machine and you get all zeros. That's positive zero. If you're coming up in the negative quadrant, the numbers get smaller and smaller and fall off the end of the machine and you eventually come up with all nines, and that is negative zero. But while this made good sense to me, it did upset a very large number of mathematicians.

Mark I had a number of built-in units. These units were in effect built-in subroutines, though of course we did not call them subroutines then. These included a sine unit logged to the base 10 and 10^x, logarithms, and exponentials. These units were built out of relays and had stored constants with which they could expand the normal series for the functions. There was only one thing wrong with them: their operation took too long. They always insisted on delivering the answer to 23 digits! Practically nobody ever needed that degree of accuracy. In addition, the sine unit persisted in computing the entire reduction of the sine, even if the programmer knew perfectly well the angle was positive and less than $\pi/4$. Now, if you had zeros in the multiplier, everything went much faster; the machine just zoomed right across. So one day I pulled the lower relays on all the coefficients and the log subroutines, and the machine ran twice as fast. We began doing this regularly, until one day we forgot to put them back in when we needed the accuracy. That put an end to that practice; from then on it was forbidden to pull any relays. Eventually the units were replaced by software equivalents.

In addition to the above-mentioned units, there were three others that were unique to Mark I. These were the interpolator units. They read paper tape just like the sequence mechanism. When you sent an argument to the interpolator, it would step through the tape until if found the argument, read the corresponding interpolation coefficients, and then apply the usual interpolation formula to give you the value you had requested. There were three of these units, and they could be loaded with different tapes to provide use of various other functions in addition to the sine, logarithms, and exponentials.

Major problems came to us one after another in those days. We were always under very heavy pressure from Washington and other places to hurry up and finish their problems. The magnitude of the problems and the adjustments in thinking the use of a computer made necessary are illustrated by a problem that was given to me. I was assigned to compute the power of a dipole which was to be towed behind a ship to sweep for mines. The authorities wanted the value of the effect of the dipole for every 20 feet in every direction for perhaps 200 feet. I calculated how much paper it was going to take to print the results to this degree of accuracy and found it would require a whole cord of wood. I reported that this would not be very useful aboard a ship, nor was it very necessary because I found the attenuation went up in a sharp loop and then trailed off; beyond a certain point it did not even change value in the eighth or ninth decimal place. So I asked if they would settle for going out only to the point where the decimal places did not change, with the result that a small book would be sufficient to show the strength of the dipole's field. Even with all Mark I's computing power you still couldn't use it thoughtlessly.

Mark I was so new, complex, and colossal that people would believe almost anything of it. Often visitors to the Center would look down the drive shaft that ran its length turning all its cams, counters, etc. and ask me what it did. I told them very seriously that we just dropped a card in at one end and it traveled down the shaft and then went up into one of the registers. They always believed me! They had little notion of how it actually worked. Just seeing the cards punched, the tapes stepped along, and hearing the counters and relays clicking away impressed most people. I recall once when a group of Admirals and other VIPs came to inspect the computer, the check register was malfunctioning and continually stopping the machine. So I stood in front of the start button keeping it depressed for the whole hour they were in the Lab. For inspections, Dick Bloch wrote a magnificent program that exercised all the machine's components without actually performing any real computations. It made the computer look impressively busy and was very reliable.

Though slow by today's standards, from the perspective of a programmer Mark I was—as I said above—remarkably similar to today's computers, and indeed, much of the experience we gained on Mark I and its successors had a great influence on later software developments. We had adopted the habit of writing pieces of code in our notebooks—pieces of code that had been checked out and were known to be correct. For instance, Dick Bloch had one routine that computed sines for positive angles less that $\pi/4$ to only ten digits. Rather than use the slow sine unit, I copied Dick's routine into my program whenever I knew it would suit my requirements. This practice ultimately allowed us to dispense with the sine, logarithm, and exponential units altogether. Both Dick and Bob Campbell had notebooks full of code, and believe me, I used them frequently! What I did not realize at the time was that I was becoming acquainted with subroutines and actually developing the possibility of building compilers. A short time later, this approach was formalized in the book by Wheeler, Wilkes, and Gill, which helped advance our thinking along these lines.

The way in which we programmed Mark I had a great deal to do with the later development of FORTRAN and COBOL and the rest of our computer languages. It is quite possible that we would not have some of the languages and other things we have today had they not been initiated on Marks I, II, and III. Looking back, the future of compilers and various higher-level languages can be traced back to those pieces of code in the Mark I programming notebooks.

This early programming experience taught me many pitfalls and shortcomings of programming that were to influence my later work. One, which forced me into the development of the compiler, was the realization that programmers cannot copy things exactly. When integrating discrete pieces of code into other programs, the programmer often had to add to all the addresses, and this led me to conclude that programmers often cannot add either. Observing the poor copying and adding habits of programmers, and recognizing that there in front of them was a machine whose whole job was to add and to copy things accurately was what led me to write a compiler in 1952. Compiling together pre-established subroutines was an idea that stemmed all the way back to those pieces of code that floated around in our notebooks in the Mark I days.

Commander Aiken was a tough taskmaster. I was sitting at my desk one day, when he came up beside me. I got to my feet real fast, to hear him say "You're going to write a book." "I can't write a book," I replied. "You're in the Navy now" was his firm reply. And so I wrote a book: the *Manual of Operation* for Mark I—the entire bible for Mark I. It contained every circuit, had samples of programming and coding, an excellent bibliography on computation, and timing charts for all the operations for various circuits. With the *Manual* you could assemble Mark I all over again, if you felt like it.

In addition, the *Manual* includes a description of how a program must be delivered. As an outline of a disciplined and thorough approach to programming, it could well be given to most programmers today. It spelled out the various parts which must be provided with a program when it is presented to be run on the computer. The first, of course, was the coding that controls the sequence mechanism. The second was the plugging instructions governing the degree of accuracy for the multiply/divide and other such units. Then the programmer had to provide instruction for the operator. These included the setting of constants, the use of any preliminary tapes to clear registers, etc. Even that far back we were trying to get programmers to annotate their coding for the poor operators who had to cope with their programs in the middle of the night when the programmer was not there. While we have made great advances in building the machines, it appears we have not had the same luck getting programmers to write comments on their code so somebody else can understand what it's all supposed to be about.

Aiken taught us discipline. He didn't consider a program complete until all the above requirements were met. Of course, Dick Bloch got away with running things without having the operations instructions complete. But he was exceptional. I know only one person who was ever able to write a complete program in ink and have it run the first time, and that was Dick Bloch. He drove us all nearly crazy by doing that. He also had another rather interesting habit. Mark I was a relay and step counter machine. Therefore, it was not too difficult to change the circuits. Every once in a while, Dick would get an idea for a new circuit that would make his problem run faster, and at night he and one of the operators would alter the circuits. The next morning, of course, when I tried my programs, they would not run. That is what happens when you let programmers meddle with hardware.

In thinking back on Mark I and Howard Aiken's achievements, I find that the computing community has totally failed to recognize many of the contributions he made, not only to hardware and to logic, but also to software. Aiken was a good teacher. He taught all of us in a rather curious way. He told us to go do something, but then left us alone and merely kept an eye on us. He led us into learning, and the experience we had with him enabled many of us to go forward in the field of computing. Above all, he gave us the impetus and the will to do it.

Howard Aiken did not only teach us about computers; he taught us to be part of the Navy. It went along with learning to run the computer. That too is a lesson that has stayed with me ever since and that I highly value.

The sense of being at the beginning of something new, the fun, the excitement of using Mark I is almost impossible to convey to people today. I contend that even today we are still only at the beginning of the industry. To compare it to the auto industry, we finally have the Model T, the computer that everyone can own. But the real beginnings, the birth of the industry go all the way back to Mark I—to all 51 feet of her, 8 feet high, and 8 feet deep, encased in gleaming stainless steel and glass design by Norman Bel Geddes. Though her output was to typewriters, tapes, or punched cards and was slow by today's standards, with 72 storage registers and three operations per second, she was able to take on some fantastic jobs: elaborate systems of partial differential equations, major statistical problems, and other tasks of similar gigantic magnitude.

Those of us who were fortunate enough to work on Mark I were extremely lucky to get in at the very beginning. The problems that we solved were big problems. Yet, had it not been for Howard Aiken, probably none of it would have gotten done, even if we had the computer. It was he who directed and led the work all those years. We all owe him a great deal.

I certainly gained confidence from my experience under Aiken. For me, Mark I will always be my favorite computer, because I knew exactly how to tell her what to do. With some of these new computers, I get a little confused. Mark I did not have any operating systems, or virtual memory, or multi-programming, or any of these things. She was simply a dedicated computer. From using Mark I, I learned the basics, what you can do on a dedicated machine which does one problem at a time. That experience gave me the confidence to forge ahead.

I have mentioned my work on compilers and languages and how those pieces of code we wrote out from our Mark I notebooks became the subroutines of the compilers. Of course when I first proposed that we should program computers in simple English, I was told it was impossible. I had a perfectly dreadful time trying to explain that a word was merely a symbol, just as much as X, Y, or Z. People just could not seem to grasp the concept that if a computer could treat the formula $X + Y = Z$, then it could also deal with an expression such as MULTIPLY WAGE BY TAX. But I persevered, and I was strengthened in my convictions by the experience and confidence I had acquired working with Mark I.

It was a tremendous accomplishment for Howard Aiken to get Mark I built, and it was also an accomplishment for the crew to run it, to keep it going, and to maintain it morning, noon, and night throughout those war years. I am very proud that the first large-scale digital computer in the United States was a Navy computer, just as I am proud to wear the uniform of the United States Navy.

Over the years I had a great deal to do with computers, and I know I have driven a large numbers of people at least partially nuts with my tenacious insistence on such ideas as programming in English. I have been given many honors in the industry. Yet each time I receive one, I confess that I have already been given the highest reward I shall ever receive—no matter how long I live, no matter how may more different jobs I may have—the privilege and the responsibility of serving proudly in the United States Navy, of working with Howard Aiken, and of programming and operating Mark I.

Reminiscences of Aiken during World War II and Later

Richard Bloch

Among my memories from the war years, there are vivid recollections of the number of occasions when I saw Howard Aiken working at his desk with the utmost of concentration on the subject matter at hand until perhaps 10 P.M., having had no dinner; he then left for home and a forbearing wife who I surmised was well accustomed to this schedule. I can recall my having to work through the night on a few of these occasions, when to my surprise Howard reappeared on the scene at 4:30 in the morning, asked how things were going, poured himself a cup of coffee, and returned to his desk full of renewed vigor, picking up with his work where he had stopped a few hours earlier.

Howard was really a member of the team—he was a "playing captain" at all times. Although he took care of administrative matters as necessary, his heart was in analysis, computation, and machine design. Many times, when he knew I was besieged with a backlog of programming for a particular problem, he would ask me which program tapes he could code to assist me; he would inspect my programming flow charts and in a short time perceive what attack I was using, how I was utilizing the storage registers, and without many further questions return to his desk, programming assiduously in my behalf. Rare was the time when any error was to be found in his work product, and I was able to combine his coding with mine with the utmost of confidence. Furthermore, his work was always fastidious, well documented, superbly organized—and always completed in a minimal amount of time.

Nor did Howard shirk some of the more menial tasks. I remember his assisting in punching lengthy sequence tapes and then, often during evening hours, verifying them with a hand-held decoder while I read off the coding from my original program pages. At the time, in the early days of the laboratory, there were only three of us on the professional staff (Howard, Bob Campbell, and I) and he knew that all of us had to pitch in on every aspect of the operation.

Mark I operated on a 24-hour basis during the war period, and the team of enlisted Navy "I-Specialists" assigned to the Harvard Computation Project were responsible for keeping the machine in operation around the clock. These men, who became very proficient at maintaining the machine operational, took eight-hour shifts in relays. One night, on the shift beginning at midnight to which one of the Navy specialists was assigned, with the machine working on one of my programs which had been running for days, a very unfortunate situation developed. It was about 4:30 in the morning; the machine was purring along smoothly with its usual distinctive clattering of relays intermingled with the sound of the clutch mechanism stopping and starting the main sequence drive; the two printing typewriters were clicking out final output at regular intervals. Apparently the monotonous regularity of these machine sounds had a most soothing effect on the Navy I-Specialist on duty at the time; he fell asleep at his operator's station! Now it appeared that not everyone had found it easy to sleep that night. In fact, one person in particular, who apparently had a bout of insomnia with too many things stirring about in his mind, decided to get dressed and get to his place of business a few hours earlier than usual on this very same morning; that person was none other than Commander Howard Aiken! Howard arrived at the laboratory just in time to catch the Navy operator blissfully snoozing. Whether the machine was running at the time or not I never did determine—not that it made too much difference. I heard about this incident a few hours later when I arrived; my problem was still running; the same operator was there finishing his shift; and he quickly confessed to me what had happened. He indicated that the Commander was not amused—which I am certain was close to the greatest understatement of all time. Within a short time, the poor fellow was replaced; I never saw him again. I also thought it best not to inquire of Aiken the location of the man's next duty station!

On one occasion, with an enormous amount of programming facing me, and with a tight schedule to get the problem running on the machine, I worked through the night and into the next day, taking time out only for a few quick meals. By nightfall, I had made good progress, but now that I saw the light at the end of the tunnel I decided to continue on through the second night without sleep; I thought I could complete the program by about noon the next day. Howard

Aiken arrived at about 8 the next morning, and upon hearing that I had not gone home for the second successive night, he had had enough. He told me to put on my jacket and that he was going to personally accompany me home. I said that wasn't necessary; I would go to Harvard Square for some breakfast and then continue home for some sleep; there was no need for his bothering to come along. Aiken would have none of it. He took me arm in arm and out of the laboratory we went. When we arrived at my quarters—I was living in Navy Bachelor Officers' Quarters on Garden Street next to the Hotel Commander—I thanked him for his concern and said I would see him the next day. He still wasn't satisfied that I meant to go to bed. He insisted on coming into my quarters, seated himself next to my bed, and told me to get my pajamas on and get under the covers; he said he was not about to leave until I was tucked in. So he waited until he was convinced that I was not going to play any games, whereupon he pulled down all the shades to shut out the morning light, and then left with the warning that I was not to show my face at the laboratory for at least 24 hours or I wouldn't be pleased with the consequences. Needless to say I wasn't about to argue the point; I was too tired to move anyway.

Grace Hopper arrived at the laboratory some time in July of 1944—the first of many Naval officers with a strong mathematical background to be assigned to duty at Harvard—in her case she had taught math at Vassar College. I remember how avidly she wanted to learn about Mark I; the computer world was new to her, as it had been earlier for all of us. Many an evening I spent tutoring Grace, either at the lab or over a late dinner at the Red Coach Grill in Harvard Square, explaining the modus operandi of Mark I and the techniques for programming the machine. As time went on, she somehow became convinced that insofar as programming was concerned, I was infallible. In later years in her talks as the Navy's grand ambassador and inter-preter of the computer industry to service personnel and civilian groups, she insisted on perpetuating the story that I was the only person who wrote his computer programs with a *pen*, no erasures ever being necessary; although this was something of an exaggeration, the story has persisted.

The main program tapes of Mark I, once punched and verified as accurate, were frequently glued end to end so as to form a continuous loop. This was done to enable the machine to repeat the execution of the program with a new set of input variables on each revolution without interruption. These tapes often were quite long and required a multiple array of guide spools arranged so as to keep the tape clear of any obstructions and feed the sequence mechanism smoothly. One bright sunny day, one of my endless program tapes was placed on the machine, when to the horror of the operator, upon inspecting the spool arrangement he found that the tape had been twisted prior to being glued. The commotion that ensued drew a crowd around the main control area; Grace Hopper had seen what had happened and shouted with unmitigated glee that Dick Bloch was trying to run a program that was written on a Möbius strip—a one-sided surface! Amid the din of laughter, I was trying to explain that the tape would run anyway—that is on its first revolution. Meanwhile, the operator, a bit red-faced, and not interested in listening to a mathematical discourse on the subject had quickly cut, untwisted, and reglued the tape; too bad, I thought, because I was about to prove my point.

During my three years at the laboratory, there were many occasions when I had the opportunity to chat with Howard Aiken in a relaxed fashion on various subjects including the world of computing present-day and in the future. One day we were discussing the work going on at the Moore School of Engineering at the University of Pennsylvania. Howard, while clearly indicating respect for the accomplishments of Eckert and Mauchly with the ENIAC machine, remarked that the ENIAC was not really a general-purpose computer; its design, he felt was unduly influenced by certain military problems it was originally intended to solve. Future computers, he averred, would hardly be designed along similar lines; furthermore, while electron tubes could clearly switch much faster than relays, he remarked that he would wager me any amount that the up-time percentage of the ENIAC did not—and never would—come close to that of Mark I. He added that until electronic devices with much greater reliability and with much less heat generation were developed, the use of the common electron tube would never become the "staple" device of future machines; and until such new electronic devices or perhaps magnetic devices evolved, he was staying with relays—the speed of electronic switching not withstanding!

On another occasion, I remember bringing up the subject of future applications of computers. At the time Howard and I were at Mark I checking the high order differences of the Bessel functions as they were being printed on one of the output typewriters. I conjectured that there was nothing to prevent computers such as Mark I and its successors from being used to advantage for business applications in the commercial world. This prospect didn't elicit a high level of enthusiasm from Howard. He remarked that in terms of mathematical complexity, applications such as business accounting were so elementary that they hardly posed a worthwhile challenge to high speed automatic computers. However he admitted that simply eliminating the discrete card-processing steps—which was the basis of the IBM business machine world at the time—and placing the entire sequence of steps under the automatic control of a single computer—hopefully eliminating the punched card altogether—would indeed be far superior in the business arena when compared with the current IBM attack. Then, pointing to the typewriters clicking away at nine characters per second, I suggested that in future computers, something would have to be done to greatly increase the speed of printing. Howard retorted: "Dick, look at these hundreds upon hundreds of pages of Bessel functions we are producing on these typewriters. As it is, no one in his lifetime will ever be able to peruse all these pages. What in the world would we do with ten times the speed? The output would never be read in ten lifetimes!" I was thinking about premium bills and bank statements; Howard was thinking about solving scientific problems.

I have been fortunate enough to have participated in the computer era from its birth a half century ago through to the present day nearly a half century later. During this period, I have been actively involved in both the hardware and software segments of the industry, beginning with Mark I, and then progressing through generation after generation of advancement. Design teams with which I have been associated across the years either in a technical or management capacity have been responsible for many advances in computer technology, a number of which have been broadly accepted by the industry generally; their impact is readily apparent in the present day environment.

The most important of these advances were the direct outgrowth of my experiences in the Harvard Computation Laboratory. For example, the trials and tribulations of checking the data and computations in Mark I programs were clearly in mind when in 1947, while working at Raytheon on a design proposal for a new computer, I developed a system for built-in error detection. I was but a few months away from Harvard, and my recollections were still vivid—I had to try to find a way of relieving the programmer from the burden of checking the machine's work and somehow enable the computer to do this automatically. The system that I devised—the weighted count check—involved several binary digits of redundancy which accompanied all data and instructions in the machine. The system was capable of not only checking all transfers of information but also all arithmetic operations as well. The parity check as we know it today was actually described for the first time in my weighted count patent as a degenerate form of weight count involving as little as a single binary digit (or "bit" as we know it today). The multiple bit parity, as eventually incorporated in this early machine, was much more powerful than simple parity in that it intercepted nearly all multiple errors as well as any single-bit error; furthermore it provided the means for checking every arithmetic operation including binary-to-decimal conversion and vice-versa. Several years later, I went further and devised a system for automatic error *correction* as well as error detection which was incorporated in a full line of commercial computers.

It is also interesting to note that the 1947 Raytheon design featured the storage of instructions in the computer's main memory—an immediate departure from the design philosophy of the early Mark machines at Harvard, but clearly influenced by the many constraints inherent in the Harvard machines as discussed earlier. This is especially understandable since the Raytheon design team—headed incidentally by Bob Campbell—consisted almost wholly of Harvard Computation Laboratory alumni.

My nearly phobic aversion to human intervention in the computer's processing of information was clearly nurtured by my experiences at Harvard. I had learned only too well the costs of such intervention both in terms of error frequency and time consumed. This was uppermost in my mind when I devised the concept of the "bar code" at Honeywell in the late 1950s. The objective here was to enable a computer at a subsequent time to read its own previously printed output, avoiding the need for any manual keyboard input or transcription in the entire processing cycle. Most importantly, the bar code avoided the relatively high error rate prevalent in computer scanning and interpretation of numerical digits or alphabetical characters. Nevertheless, the bar code was designed to provide full error detection, even though the error rate proved to be negligible in practice. Subsequent versions of this bar code are in worldwide use today in myriad forms of transaction processing.

The impact of my Harvard experience upon my entire subsequent professional career is indisputable. I owe an unending debt of gratitude to Howard Aiken in particular for introducing me to computer-dom at its inception, and of even greater importance for being a superb mentor and a source of inspiration during my entire period of association with him at Harvard. For many years after the Harvard Laboratory experience, I had only occasional contact with Howard. However, in 1972 I met with him on several occasions at his home in Fort Lauderdale to discuss the possibility of launching some new computer ventures he had in mind. One was the possibility of producing a cryptographic device for enciphering and deciphering computer-derived, stored, or transmitted data of a critical nature. He had recently obtained a systems patent embodying a number-theoretical technique for generating a pseudo-random sequence of digits of huge periodicity. The arithmetic speeds required, considering the capabilities of computer circuits of that day, it appeared, might make such a device very expensive for general usage. However, the attack was very clever and I thought that it was worthy of further study. Unfortunately, it was shortly after those meetings that Howard took ill and further work was postponed; unfortunately his death came a few months later.

To the end, his facility with, and love for applied mathematics—especially computational mathematics—was of the highest level. He was one of the great pioneers of the century and undeniably in my mind the true father of the computer industry.

As a freshman in 1947, I fended off Harry Rowe Mimno's well-intentioned advice that I needed a solid grounding in rotating electrical machinery. This experience made me so shy of Harry that it was not until years later, when the dean appointed him to the committee to write Aiken's "memorial minute,"[2] that I realized how important Harry had been in Aiken's career.

Having often heard that Aiken ate undergraduates for breakfast, I sought out as my departmental adviser Philippe Emmanuel LeCorbeiller, like me a French-speaking refugee from Hitler's European rampage. Phil LeCorbeiller for many years remained a wise counselor and protector when I was faced with some unfathomable Aiken irascibility. Phil's main initial advice was to look for the right moment to approach Aiken. Meanwhile, with Winston Churchill's "Iron Curtain" speech fresh in mind, I started studying Russian for self-preservation, along with the calculus and such for my major and French literature for fun.

My thirst for computers unslaked, I wangled a job at IBM's (and Wallace Eckert's) Thomas J. Watson Laboratory at Columbia University for the summer of 1949, and I returned there in the summer of 1950. My boss both summers was Bob Walker, an alumnus of MIT's wartime Radiation Laboratory who, inspired by Vannevar Bush's analog differential analyzer, had built an analog machine for solving linear equations. He had me work on that the first summer. At the beginning of the second summer, he tossed on my bench a box full of RCA's then-new Type A transistors and asked me to see if I could make digital counters from them. (I could and did. But the transistors were so unstable that I had to put a variable potentiometer in every critical circuit in order to keep adjusting everything. This led me to recommend staying away from transistors for computer use, a tactically correct inference. Of course, I did not know at that moment that I would be dead wrong strategically. The Type A (point-contact) transistor was about to be cast into the dustbin of history by new fabrication techniques that have by now put transistors in the zillions among the most stable, reliable, and economical artifacts ever.) I had ample time to hobnob with Byron Havens and his assistants, who were building

2. Besides Harry Mimno and me, the other members were Wassily Leontief and I. Bernard Cohen. I was the chairman. The "memorial minute" can be found in the records of the 21 May 1974 meeting of the Harvard Faculty of Arts and Sciences.

the vacuum-tube based Naval Ordnance Research Calculator (one of IBM's first electronic digital machines), and to schmooze with the young and hyper-ebullient Herb Grosch. In that milieu, I took some kidding as a Harvard student and learned of the mythic Aiken as portrayed in IBM, MIT, and Columbia demonology—not a reassuring experience.

I also got acquainted with "Old Tom" Watson and Jeannette, his wife, who in those days still came to the lab picnics. Looking back from 1992 I cannot remember whether it was then or much later that their son-in-law John McPherson, who was then somewhere up the Watson Lab's reporting hierarchy, told me of his presence at a meeting where Aiken asked Old Tom for support for what was to become the Mark I/ASCC. John said that he had counseled his father-in-law against supporting Aiken. I asked him why. "Because," John told me, "my engineer's soul rebelled against the crazy idea of building a machine 90 percent of which would be idle 90 percent of the time." In contrast to McPherson's outlook, think about the pads of paper each of us owns, the books on our shelves, the telephones in our offices, and the cars we drive, all of them unused most of the time: once things get convenient enough and cheap enough, the effectiveness of abundance (erstwhile wastefulness) overwhelms the efficiency of high load factor (latter-day miserliness). I think of this each time I walk past Aiken's relic (a section of Mark I reassembled for display) to my desk, where a Compaq 386 sits idling. At its peak, Mark I had about 100 storage elements. The Compaq has 10 megabytes of random-access memory and a gigabyte of hard-disk memory.

Thus, I was steeped in the odor of IBM when I first met Professor Aiken. My confession of this got not the slightest rise out of him, and as far as I ever could tell he never held it against me. Following Professor Le Corbeiller's advice, I had chosen the occasion carefully. It was the academic year 1949–50, after my first IBM summer. I had been among eight Harvard juniors elected to Phi Beta Kappa, and the full undergraduate membership had picked me as their first marshal— a dubious honor that carried with it, along with unannounced and mind-numbing clerical chores, the responsibility for the smooth functioning of the annual lunch with the newly elected honorary members from the faculty. Aiken happened to be among them, and I vividly remember my glee at controlling the seating arrangements. And so it was that I found myself and Professor Aiken, both with martinis in hand, side by side at the head table. In the era of the cult of the dry martini, a cult that Aiken played for all it was worth, this turned out to have been exactly the right thing to do. In later years, when an exasperated Aiken would occasionally threaten to cut me up into little pieces and feed me to myself or the like, I recalled that I had brought it all on myself.

When I first walked in, Harvard's Computation Laboratory was in the process of shipping Mark III to the Naval Proving Ground at Dahlgren and was starting the construction of Mark IV, but Mark I was still operating and even growing. I had a part-time paid job helping Gerrit Blaauw, Johnny Harr, and others finish the vacuum-tube/magnetic delay-line Mark IV. Of that job, I mostly remember dreary hours in the basement lacing miles of wires into cables.

But I also found myself welcome at the famous coffee hours and talking to graduate students, other members of the staff of the Comp Lab, people from everywhere at Harvard, and people apparently from nowhere. Among them were An Wang, who was then finishing a faster multiplier for Mark I (and thinking, presumably, of taking out the patents that would help launch Wang Laboratories) and the Reverend Ralph Ellison (whom I assisted in eventually banging out of Mark IV a concordance of the Bible, a feat that in retrospect seems comparable to Archy the cockroach eking out *The Life and Times of Archy and Mehitabel* one lower-case letter at a time by jumping head-first onto each key). And I met Professor Wassily Leontief, who was mulling over using Mark I to do, with Ken Iverson's assistance, the input-output calculations that helped toward his eventual Nobel Prize in economics. And somewhere along the line Aiken introduced me to Bob Anthony, the accounting guru at the Harvard Business School, with whose graduate student Chuck Christenson (now Business School colleague Charles Christenson) we eventually squeezed out of Mark IV the first published tables of present values, which were eventually to revolutionize business financial practices and which still reverberate to this day in the financial service industry's destabilizing "derivative products."

Probably I am not doing justice to others who were there between 1949 and 1961, but there is more of that elsewhere in this book. It was heady stuff, all right. With all this going on and with my own IBM and other windows on the world, I never felt the claustrophobia or parochialism that others report elsewhere in this book. But I had my own perceptions of a dark side of Aiken. I disliked his speaking derisively of the Walsh-Healy Act, which afforded certain protections to labor. Many of us rolled our eyes when we heard him refer to a black part-time janitor as "Old Smoke." But I did not speak up, and I heard no one else confront him about it; this was, it should be remembered, "the silent fifties," an interregnum between the "y'all and shut mah mouth land" attitudes that my fellow graduate student Tom Lehrer lampooned in his song *Dixie* and the protest movements of the 1960s that segued into the occasional excesses of political correctness of the 1990s.

In any case, my joy lay in the keys to the laboratory, which Aiken literally gave me. Slavey was slavey and student was student, and Aiken treated me in each role as his view of the role called for. Here I was, a junior in Harvard College, able to walk at will, day or night, into this citadel of high technology and learning.

There was an obligation, of course. After the Phi Beta Kappa conversation, where Aiken quizzed me on my background, we had a meeting where his evaluation of that background had clearly fixed in his mind the road on which he would set me, a choice thoroughly consistent with the observation (recurring in this book) that Aiken prized computers only for what they could *do* and not for niceties of high technology. Whatever Aiken was, he was not what we would today call a "techie." He was, of course, a superb operator on a stage of mind-boggling scope. He sat me down and showed me a letter from Warren Weaver,[3] then head of the Rockefeller Foundation. In that letter, Weaver speculated on the possibility of using computers to translate languages, reasoning by analogy from what he knew (in 1949) about wartime use of computers for cryptography. "See what you can do with your Russian," Aiken told me.

And so, throughout the rest of my junior year and all through my senior year in Harvard College, I began working on what in 1954 became my doctoral thesis, A Study for the Design of an Automatic Dictionary. Aiken opened for me the doors of I. A. Richards, then fresh from inventing Basic English, and Joshua Whatmough, the chairman of Harvard's linguistics department. This was enough to launch me into collegial circles at Harvard and at MIT. Aiken also introduced me to Air Force people at the Rome (N.Y.) Air Development Center and

3. It is only from Cohen's work that I have learned that this was but an episode in a longer correspondence between them; I did learn soon thereafter that the particular letter had been widely circulated in certain circles and had inspired some research in machine translation.

at the Foreign Technology Division at Wright-Patterson Air Force Base in Dayton—potential users for whatever might come of my work. This set the stage for my lifelong involvement in national security affairs. Those who saw Aiken as parochial saw only one facet of him. At least to me, he was most generous in sharing some (but surely not all[4]) of his many windows on the world.

"Ask and ye shall be given" seemed to be Aiken's principle, so long as you produced. I got minimal guidance on details. I remember working on something, perhaps a thesis draft, on which I wanted his opinion. He hoisted up his forbidding pince-nez on his big face with the Mephistophelean eyebrows and the sardonic grin, flicked through the pages, and tossed it back on my desk. "It stinks," he threw over his shoulder as he turned to stalk out.

On the basis of Le Corbeiller's advice and in spite of Aiken's "What do you want to do that for?" I applied for and got a fellowship that allowed me to spend my first postgraduate year at the other Cambridge, working for Maurice Wilkes on his newly operational EDSAC computer at the Cambridge University Mathematical Laboratory. Wilkes had no use for language translation and harbored the mixed feelings about Aiken that he describes in this book.

The linguistics professor at Cambridge was then a man whom Joshua Whatmough described to me with his customary charity as "one who speaks many languages but says nothing in any of them." So I followed Wilkes's suggestion that I address whether computers might be made to learn. My nights on the EDSAC were often shared with John Kendrew, who came across the Cavendish courtyard to do the calculations of the structure of myoglobin that were part of work for which he later got his Nobel Prize. My colleague Hank Houtthaker (now in Harvard's Economics Department) was also a frequent competitor for machine time. Alan Turing and others became familiars at meetings of the Ratio Club. My first published paper, "Programming a Digital Computer to Learn" (*Philosophical Magazine* 7, no. 43, 1952: 1243–1263), came out of this ferment. It was one of the earliest in artificial intelligence and one basis for my recurrent criticisms of the excesses of the artificial intelligentsia. I was invited to present the paper after my return to Harvard, and I did so under the title "Simple Learning by a Digital Computer" at the meeting of the Association for

4. For instance, I had no part in his dealings with the National Security Agency.

Computing Machinery in Toronto in 1952. I recall announcing my intent to do this presentation in Aiken's presence and eliciting a shrug of the shoulders. I think he regarded my whole Cambridge stay as some aberration of a prodigal son. Though he clearly hadn't been for it, neither did he give me any evidence of holding it against me. In any event, disillusioned by the prospects for artificial intelligence, I returned to my work on automatic language translation. Forty years later, techniques like those I developed at Aiken's suggestion have at last become affordable not only for experimental or upscale institutional users but even in the much broader consumer electronics market.

While I worked on my thesis, with its dubious applicability,[5] Aiken roped me into his many exploratory incursions into business data processing. We worked with the Hood Milk Company on a scheme for using computers to pay milk producers. We had a good working relationship with folks at the American Gas Association and the Edison Electric Institute, working on (among other things) automating meter reading. I traveled with Aiken to Lockheed's Missile and Space Division near San Francisco to meet with Howard Resnikoff and other language-translation people. Our long-running contract with AT&T's Bell Laboratories influenced not only us but also the people who eventually brought digital switching to the world's telephones and opened up my lifelong friendship with Deming Lewis and William O. Baker. All these ties Aiken somehow brought to life and kept going, with us wrapped up in them, until he retired and beyond.

My field was too new-fangled for routine faculty appointments. Aiken had no budget for me, so I spent my first postdoctoral year (1954–1955) nominally as an "instructor" but with my pay coming from a National Science Foundation postdoctoral fellowship (which I learned I had gotten while I was away, jobless, on my honeymoon). The job market then was poor. With the Soviet Union's startling launch of its Sputnik satellite still three years in the future, the boom years for American science and technology were yet to come. For 1955–1956 Aiken and Whatmough somehow managed to get me appointed as a "split instructor" in linguistics and applied mathematics. For the summer of 1955, however, Aiken sold me to a bank. We

5. Jackie Sill and I were the only ones in the whole world fluent in a two-digit code for Russian letters. That code was our only means for input and output in those days.

had made numerous trips to the Philadelphia National Bank, where a farsighted assistant vice-president named Lew Nungesser had formed a vision of electronic banking. We were treated well: Aiken insisted that scholars deserved it. We saw Robin Roberts pitch for the Phillies from the bank's box, for example.

Ultimately, Aiken persuaded the bank that they needed a resident Ph.D. to help them along, and I was it. There was a protracted negotiation, mostly conducted by Aiken on my behalf, regarding where a Ph.D. would fit into the social fabric of the bank. Aiken clearly won on that score, and I found myself not only eating lunch with the top officers in their private dining room but also holding a coveted key to the executive john. In real life, I lived in the YMCA on Arch Street. Later, as a consultant to Arthur D. Little, Inc., from 1956 to 1980, I developed that Aiken jumpstart into expertise in bank automation, which continues as a research interest of mine to this very day. Looking at the big picture was instilled in me by Aiken's insistent example. I think Aiken would relish the following juxtaposition; I do:

It is . . . surprising that, on the whole, most banks and bankers continue to regard automation as a mere technical matter for the individual bank, whose only objective and only possible consequence is to cut the costs of doing routine work or to increase the volume of routine work that can be handled for a given cost. In reality, automation affects not the mere mechanics of banking, but the very foundations of banking; not the individual bank, but banking systems and the national and international economies in which they are imbedded." (Anthony G. Oettinger, "The Coming Revolution in Banking," in *Proceedings: National Automation Conference 1964* (American Bankers Association, 1964), p. 38)

In retrospect it appears that the transition from manual, paper-based systems to electronics may not have been managed by those sufficiently sensitive to credit and risk exposures. In both the private and the public sectors, low-cost, high-speed, highly efficient systems were built that delivered what was requested: speed at low cost. Only after the fact did we all become aware that the financial systems were at risk with serious implications for world markets. . . . We must all guard against a situation in which the designers of financial strategies lack the experience to evaluate the attendant risks and their experienced senior managers are too embarrassed to admit that they do not understand the new strategies." (Alan Greenspan, "International Financial Integration," remarks before Federation of Bankers Associations of Japan, October 14, 1992, pp. 3 and 16 of typescript)

Aiken's retirement in 1961 seems not to have been entirely voluntary. He hinted that, although he could not be forced out of a tenured professorial appointment, his administrative appointment as Director of the Harvard Computation Laboratory was at the discretion of the Dean of the Faculty of Arts and Sciences, then McGeorge Bundy. (Bundy later became National Security Adviser to Presidents Kennedy and Johnson, and then president of the Ford Foundation.) Aiken also said that it was time for him to make money, and that he thought that he should be able to do so even on half his cylinders. This he went and did with great gusto and apparent success.

For two or three years preceding his retirement, Aiken cut me off. To this day I have been unable to figure out why for sure. I thought at first that I had offended him in some way, but I could not learn how, especially since he hardly more than grunted at me on anything, let alone on so delicate a subject. I made no secret of my plight, especially when querying associates for clues about whatever offense I might have given. A light began to dawn when Bundy called me into his office in 1960, the year before he too left Harvard, to tell me that I had been promoted to a tenured associate professorship. The meeting took an odd turn for me when Bundy quizzed me about allegations that the Comp Lab was an authoritarian kind of place where people referred to the Director as "Boss"—apparently an unacceptably unacademic hierarchical expletive in Mac's lexicon. I can't recall what I replied, only that it didn't seem to soothe Bundy. But I do remember thinking as I walked out that, authoritarian for authoritarian, I'd take the Boss anytime—although he *was* hard to take at 8:00 A.M., when he made his daily rounds and popped into graduate students' offices with his damn cheery query about the progress of their work: "What'cha got for me today?"

If how many nicknames you have indexes affection, Aiken was a 10. The terms I remember us using at the Comp Lab are "Commander" (to his face), "Professor Aiken" (in front of strangers), and "The Boss," "The Old Man," "The Tall Man," and much worse (behind his back). Only friendly visiting peers, like the military people or Deming Lewis from Bell Labs, called him "Howard"; for me, even after his retirement it was still "Commander" at first and "Howard" only in his last years.

I believe that in snubbing me the Commander was simply acting Machiavellian. Knowing me well enough to predict that I would whine publicly and thereby disassociate myself from him in all innocence, he set me up for a tenure recommendation that he thought would otherwise be turned down on the suspicion that I was his boy. Some time after his retirement, I put the question to the Commander, who by then had reverted to speaking with me and who eventually became a friend. All I got beneath the Mephistophelean eyes and eyebrows was that inimitable grin—more Puck than Mephistopheles on this occasion, but inscrutable nonetheless. Considering that it was the turn of the 1940s when Harvard's President James Bryant Conant told Aiken that he would not get tenure if he persisted in his quest for the computer, and considering that it was the turn of the 1990s when Harvard's Faculty of Arts and Sciences found it timely to create a Standing Committee to oversee its use of computers, I think my conjecture has merit. Thus, in my eyes Aiken is a hero of academic guerrilla warfare as well as a titan of high tech.

On 6 October 1961, Aiken's friends rallied around him at a party organized by Jackie Sanborn Sill. We had solicited all of them for contributions, and Jackie had had a Rockport silversmith friend sculpt a sterling silver replica of Mark I.

Meanwhile, although secure in my own tenure, I apparently did not feel comfortable enough until later to look after ensuring Aiken's recognition as a prophet in his own country. Even then, I did not take the initiative. There is in my file a letter, dated 24 March 1964, from Isaac L. Auerbach, then chairman of the now-defunct American Federation of Information Processing Societies, that says: "Many thanks for your letter making the formal nomination of Howard Aiken. Apparently, however, you were never given the form that was designed for this purpose. . . . In addition, may I ask that you send me four additional copies of the attachments to your letter to me." Since Auerbach copied his letter to Samuel Alexander, Alston Householder, John McPherson, and Jerry Noe, I presume that group had picked me to formalize their intent. At some point before March 24, I had sent Auerbach a letter that began: "I am more than delighted both personally and as a member of the computing profession to have been given this opportunity to make the formal nomination of Howard Aiken as the first recipient of the Harry Goode Memorial Award."[6]

Things had more or less thawed between Aiken and me by 1964, when he asked me to join him at a dinner to be held on 21 October. He was to be awarded the Franklin Institute's John Price Wetherill Medal, and he wanted me to accept the medal for him. Here was Aiken at his theatrical finest, and with that recurrent tinge of crudeness. He had always loved to drop a reference to some king here or there. In

6. My file copy of that letter is undated, but I can't imagine that I would have written it after March 24.

this case, winking at me over cocktails, he let everyone know that he was off right after dinner to Madrid to see the king and get a medal. Came dessert, he marched out with a great flourish. The bland official record of the event has the chairman recognize Professor Y. H. Ku to say: "Mr. President, I present, in absentia, Howard Hathaway Aiken for an award."[7] After Ku read the citation, the chairman, Dr. LePage, turned to me and said: "In the absence of Professor Aiken, and at his request, I call upon Professor Anthony Oettinger, of Harvard University, who has kindly consented to accept the award in his absence."

On 5 November 1964, emboldened by support from Dean of Applied Sciences Harvey Brooks, I wrote to Harvard's president, Nathan M. Pusey, apprising him of the aforementioned honors and adding: "I think that enough time has passed to enable this University to consider dispassionately how it might honor its own graduate and Professor Emeritus. I feel that it would be most appropriate first to rename the Computation Laboratory the Aiken Computation Laboratory at an appropriate ceremony following the completion of the new third floor presently under construction and second by awarding Aiken an honorary degree." In a letter dated the very next day, Pusey replied: "I shall forward the material you sent me to the Corporation's committee on honorary degrees for their consideration and in the meanwhile shall keep in mind your suggestion about the Laboratory." My copy of that letter has a notation indicating that we copied it for Dean Brooks on 12 November 1964.

With amazing speed, at the 16 November 1964 meeting of the President and Fellows of Harvard College it was "voted, in recognition of the outstanding contributions to science and to the University of Professor Emeritus Howard H. Aiken, that the Computation Laboratory on Holmes Field shall be designated hereafter the Howard Hathaway Aiken Computation Laboratory."[8] I seem to have repressed all memory of what we did behind the scenes to pull this off, with the exception of a very clear memory of two key actors. At the time, Charles A. Coolidge was one of the senior members (perhaps the most senior) of Harvard's main governing board, the Corporation (formally the President and Fellows of Harvard College). It was, for instance, Coolidge who was acting president when Pusey took his year off in

7. *Journal of the Franklin Institute* 278, no. 6 (December 1964): 437–438.

8. Excerpt from the record, signed by Corporation Secretary David W. Bailey, addressed to me with copies to Deans Ford and Trottenberg.

India. The astute reader will find a clue to the Corporation's approval of the name change and the honorary degree in the list of Aiken's graduate students at the end of this volume.

By June 1965, Harvard could award Aiken his honorary degree in a mood relaxed enough to add humor to the citation: "It shall be the enduring fame of this distinguished teacher and scientist to have made the first mark in a revolutionary new field of technology." The program delicately explained that Aiken had "conceived and helped to develop the Automatic Sequence Controlled Calculator known as Mark I." I was relaxed enough to have written to President Pusey: "May I congratulate you on the magnificent pun in Howard Aiken's citation. All those in the know enjoyed it very much."[9]

Aiken, on his part, forgave and perhaps forgot somewhere along the line. Queries his widow put to me shortly after his death led me to inquire into the terms of his will. My records indicate that he bequeathed whatever was to remain of his capital after her death to the President and Fellows of Harvard for the general purposes of the university.

No one else has captured the fine grain of Aiken's human side as well as Jacquelin Sanborn Sill in the eulogy she read at his memorial service (held in Cambridge on 7 May 1973), which was entitled "A Non-Academic Tribute to Howard Aiken." Jackie reminded us that Aiken could be a "patient, considerate, and compassionate guide" to his students in a way that contradicted his often forbidding appearance, language, and behavior. She used three adjectives to describe him: "human," "humane," and "patriotic." Recalling Aiken's "great humanity and his concern for the welfare and careers of his students and associates," she remarked that it was such aspects of his personality that caused many of his students, colleagues, friends, and associates "to remember him with such deep affection."

9. From my personal file copy of a letter dated 18 June 1965.

A View from Overseas

Maurice Wilkes

I first met Howard Aiken in August 1946. I had just come away from the famous course on computers at the Moore School in Philadelphia, where I had been greatly stimulated by meeting Eckert and Mauchly and hearing about their ideas. I had also heard a lot about the Harvard Mark I computer to be put into service. I was anxious to see it and to meet the man who had shown such remarkable vision in the late 1930s.

The Mark I was then installed in the basement of the Physics Research Laboratory. Aiken himself showed me round the machine, which was busy computing Bessel functions. It was very impressive. It was running, Aiken told me, twenty-four hours a day, seven days a week. If one controlled such a scarce resource, he said it was one's simple duty to see that it was fully utilized. One thing I particularly remember is that he said that they had caught the printer out making mistakes and were fitting checking circuits; he had discovered, as Babbage would have done, that machines sometimes make mistakes.

Mark II, a relay machine, was under construction at the time, and I saw that too. It was not obvious at that time that there was little future in relay machines and there were some very interesting things about Mark II. In particular, there were the relays which were designed on new principles: a square magnetic circuit, silver contacts, latching coils. Moreover they were designed to plug in. It was a new approach to an old problem and for that reason very interesting.

The next time I saw Aiken was in November of the same year, 1946, when he came on a brief visit to England. It was just at the time that the Mark I was being moved to the newly built Computation Laboratory, and Aiken had given instructions that, when it was working in its new location, a transatlantic telephone call should be put through to him in England to announce the fact. It was typical of the flamboyant side of Aiken's character that he should have given that instruction, since in those days only financial tycoons were in the habit of making transatlantic calls.

Aiken had with him a number of slides of the Mark I that he was very anxious to show us. In fact he gave us a brief but polished presentation. He stressed the computation of tables of mathematical functions, such as Bessel functions, as being a major application of automatic computers and said that one of the problems he foresaw was that of getting all the tables published. This seemed to us a somewhat narrow view of computation, but it is fair to add that he did mention a number of other computations that had been done on the Mark I.

As we walked back from seeing the slides, Aiken turned to me rather brusquely, as he was apt to do, and said: "You are not committed to an ultrasonic memory, are you?" I answered equally brusquely—I admit to being a bit afraid of Aiken at that time—that I was not committed to anything. This brought the discussion to an abrupt end. In fact, in Cambridge, we were only just starting our experiments with mercury tanks and we did not succeed in demonstrating the circulation of pulses until January of the following year. To that extent it was quite true that we were not committed.

I am sorry now that I did not ask Aiken to explain what his position was. It could not have made any difference to what happened, since at that moment of time there were few alternatives open to anyone wishing to build a memory that would operate at electronic speeds, and as far as we were concerned the ultrasonic approach was by far the most attractive. However, Aiken caught me on a sensitive spot, since developing a successful memory was critical to our whole enterprise and I was only too well aware of the risks.

I did not meet Aiken again until August 1950 when I paid my second visit to the United States. I naturally made a special point of going to Harvard. Mark I was working in the new building, and Mark II had gone to Dahlgren where I had seen it a little earlier on the same trip. In the work area below the Mark I, Mark III was being put together.

The bustle in the Computation Laboratory was infectious, but it was a said fact that Aiken was no longer in a position of leadership. By that I mean that the leaders of the other groups who were building the first generation of electronic computers paid little attention to what was going on at Harvard. I knew this because I was in touch with all of them: Alexander at NBS; Huskey, also with NBS but on the West Coast; Williams at Manchester; Colebrook at NPL; von Neumann and Bigelow at the Institute for Advanced Study, Princeton; and, of course,

Eckert and Mauchly. All these groups had every reason to be confident that they were working along the right lines; their machines were either working or sufficiently advanced for there to be little doubt of success. In particular, the EDSAC at Cambridge had begun to work in May 1949 and had proved the viability of the ultrasonic memory.

Not only had Aiken isolated himself from the new ideas of machine structure that are associated with the term stored program computer, but he was technologically backward. It was partly a question of an age gap. Very few people who were much older—say five years older—than I was had green fingers for electronic circuits. It was a young man's game. There are areas in integrated circuits of which the same thing could be said today. I had to come to computer design by way of ionospheric research and wartime radar. Others had different immediate backgrounds, but we were all used to wide band widths and short pulses, and we saw the possibility of achieving very high speeds with elegant economy of equipment. Aiken's technological aims seemed pedestrian by contrast.

Mark III and Mark IV no doubt filled a gap and did useful work for their sponsors in the period that elapsed before stored program computers became generally available, but they represented the end of a period. I did not find myself drawn to study them in detail. It was different with Mark I and Mark II, and indeed these machines remain of importance in the history of the period before the modern electronic computer had established itself. I analyzed them in some detail in my book *Automatic Digital Computers*, published in 1956, but written in part earlier.

Aiken considered that the computer field was about to stabilize. I remember his saying so in a lecture delivered before Mark IV was completed, in which he announced that no more machines would be built at Harvard. In fact, as we know, the field continued to rush ahead like the Mississippi in perpetual flood, and it still has not stabilized.

I came to feel that I had little common ground with Aiken and that arguing with him on technical matters would be fruitless. However, during the visit in September 1950 that I have just been describing, he egged me on by making some unkind remarks about the binary system, which he implied was a foolish system to adopt. I disagreed, saying that it seemed to me clear that you could build a much simpler binary computer than a decimal one. He replied that perhaps at the moment you could, but that when more was known about codes for expressing decimal digits in binary form, the situation would change. He was referring to work then going on in the Computation

Laboratory (see "Synthesis of Electronic Computing and Control Circuits," *Annals of the Computation Laboratory* 27, 1951). I could not agree and said so. I do not remember the details of the subsequent argument, but I know that we both enjoyed it. At all events, he invited me to his home for dinner. Bowman of the Mellon Institute and Jones from Wisconsin were there also. It was a delightful evening and I felt elated at being on such good terms with Aiken who was after all my senior and a distinguished figure.

One tends to think of Aiken and the Computation Laboratory as though they were the same thing and it is true that he dominated the place. Publications like the one just mentioned were ascribed to the staff of the Computation Laboratory and there was a list of names in alphabetical order with Aiken's first. Peter Elias, who later became my colleague at MIT, told me that they all used Boolean algebra but Aiken did not like it and would not allow it to appear in publications. Instead they had to use an inferior system based on ordinary algebra. It is only fair to add that these remarks apply to the very early days when they were perhaps a hangover from the Naval regime under which the Laboratory had operated during the war. Oettinger assures me that in his time it was like any other Harvard laboratory as far as the faculty were concerned.

Nevertheless, Aiken continued to dominate the Laboratory. It was no doubt a stimulating environment to work in. I felt, while listening to talks given by Harvard men, during the Pioneer Day devoted to Harvard at the National Computer Conference held in May 1983, that some of them did not even then appreciate the extent to which they had been sheltered from what was going on in the wide world. However, Aiken gave them a training that was sound in all essentials and, when they moved out, they made their mark. He did not, like other dominating men, tend to surround himself with nonentities or yes men. He was a judge of people and he chose for the staff of the Computation Laboratory men and women of outstanding ability.

Aiken was a man of action, rather than an intellectual; he was a superb organizer and a leader of men; he knew exactly what he wanted and how to get support for it. He was not afraid to be a controversial figure. He never let his sponsors down. He may have appeared bluff and uncompromising but, as I found that evening when he took me to his home, he could also be a delightful companion. Above all, he was, in terms of practical achievement, the first of the computer pioneers.

IBM AUTOMATIC SEQUENCE CONTROLLED CALCULATOR

60 CONSTANTS   72 STORAGE COUNTERS   MULTIPLY - DIVIDE UNIT   FUNCTIONAL COUNTERS   INTERPOLATORS - 1. 2. 3.   SEQUENCE CONTROL   TYPEWRITERS - CARD FEEDS - CARD PUNCH

PRESENTED TO HARVARD UNIVERSITY BY INTERNATIONAL BUSINESS MACHINES CORPORATION
Through the Courtesy of
MR. THOMAS J. WATSON, President.

DESIGNED BY ~
Condr. Howard Aiken, USNR
Mr. C. D. Lake, IBM Corp.
Mr. F. E. Hamilton, IBM Corp.
Mr. B.M. Durfee, IBM Corp.

AUG. 7, 1944.

The Automatic Sequence Controlled Calculator, August 1944.

The ASCC/Mark I in IBM's Endicott Engineering Laboratory, November 1943. Courtesy IBM Archives.

A Harvard delegation meets with IBM officers and engineers, December 1943. Seated: Charles A. Kirk (vice-president in charge of manufacturing, IBM Endicott), Howard H. Aiken, F. W. Nichol (Vice-President and General Manager, IBM), James B. Conant (President, Harvard), Clair D. Lake (IBM engineer), Harald M. Westergaard (Professor of Civil Engineering and Dean of Graduate School of Engineering, Harvard). Standing: R. H. Austin (Special Representative for IBM Sales Division No. 2, Cincinnati), W. Wallace McDowell (Manager of Engineering, IBM), John C. McPherson (Director of Engineering, IBM), Harry Rowe Mimno (Associate Professor of Physics and Communication Engineering, Harvard), Harlow Shapley (Paine Professor of Practical Astronomy and Director of Harvard College Observatory), Frank E. Hamilton (engineer, IBM), Percy William Bridgman (Hollis Professor of Mathematics and Natural Philosophy, Harvard), Benjamin M. Durfee (engineer, IBM), Edwin C. Kemble (Professor of Physics and Chairman of Department of Physics, Harvard), Emory Leon Chaffee (Rumford Professor of Physics, Professor of Physics and Communication Engineering, and Director of Cruft Laboratory, Harvard), John C. Baker (Professor and Associate Dean of Graduate School of Business Administration, Harvard), Theodore H. Brown (Professor, Graduate School of Business Administration, Harvard), Edward M. Douglas (Director of Sales Promotion, IBM), Donald R. Piatt (engineer, IBM), Dwayne Orton (Director of Education, IBM), Thomas E. Clemmon (IBM Sales School). Courtesy IBM Archives.

The staff of the ASCC/Mark I, August 1944. Seated: Lt. (jg) Richard M. Bloch, USNR; Lt. Comdr. Hubert A. Arnold, USNR; Comdr. Howard Aiken, USNR; Lt. Grace M. Hopper, USNR; Lt. (jg) Robert V. D. Campbell, USNR. Standing: C. Bissel, Delo L. Calvin, Frank L. Verdunk, Hubert M. Livingston, Durward R. White.

Thomas J. Watson Sr. in front of the ASCC/Mark I with the four Navy enlisted men assigned to the computer project. The "I" on the sleeve of the uniform indicates specialist experience with IBM accounting machines.

Massachusetts governor Leverett Saltonstall, IBM president Thomas J. Watson Sr., and Harvard president James Bryant Conant in Harvard's University Hall at the dedication of the ASCC/Mark I.

The ASCC/Mark I at Harvard, July 1944, fully assembled but before the protective sheath was installed. The four devices seen in the right half of this photograph are three function tape readers and a program tape reader.

A sequence control mechanism for the ASCC/Mark I.

A storage counter or accumulator unit for the ASCC/Mark I, October 1945. According to the Manual of Operation: "Each counter wheel is an electro-mechanical assembly consisting of the following major components. . . . (1) a commutator mounted in a molded plastic part, B and J, commonly called a 'molding,' having a half slip ring and ten segmental contacts numbered 0 through 9; (2) a pair of stranded wire brushes, C and F, which rotate to connect one of the contact segments with the commutator half slip ring; (3) a magnetically controlled clutch, D, which engages to connect the continuously rotating gear, A, with the sleeve on which the rotating brushes are mounted; (4) a ten's carry contact which operates in conjunction with an external relay circuit to provide carry to the counter wheel in the next higher columnar position when the counter wheel under consideration passes through ten; (5) a nine's carry contact which also operates in conjunction with an external relay circuit to provide carry to the next higher counter wheel when the wheel under consideration stands on nine and the next lower wheel has passed through ten; (6) and finally, a socket, G and K, by which the counter assembly may be jack-connected to the calculator wiring."

An assembled counter wheel (upper right) and three types of pluggable relays used in the ASCC/Mark I. Courtesy IBM Archives.

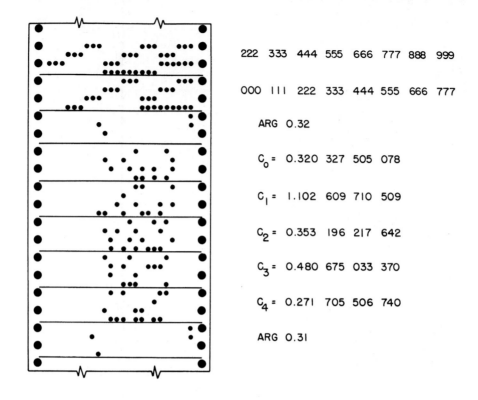

$$222 \quad 333 \quad 444 \quad 555 \quad 666 \quad 777 \quad 888 \quad 999$$

$$000 \quad 111 \quad 222 \quad 333 \quad 444 \quad 555 \quad 666 \quad 777$$

ARG 0.32

$C_0 = 0.320 \; 327 \; 505 \; 078$

$C_1 = 1.102 \; 609 \; 710 \; 509$

$C_2 = 0.353 \; 196 \; 217 \; 642$

$C_3 = 0.480 \; 675 \; 033 \; 370$

$C_4 = 0.271 \; 705 \; 506 \; 740$

ARG 0.31

A punched value tape for input of a function, as used with the ASCC/Mark I, October 1945.

The construction and design crew of Mark II, October 1946. Robert Campbell, wearing a vest, stands at the center of the second row. Seated in front of Campbell, also wearing a vest, is Frederick Miller, who was in charge of the overall construction.

Wiring Mark II, 1946. At the extreme left is William Porter, who was in charge of the construction crew.

Wiring within a relay cubicle for Mark II, September 1946.

Three new types of relays specially designed for Mark II, January 1947. Center: a latch relay for permanent storage. Left and right: single-coil relays.

A high-speed tape-reading mechanism for Mark II, January 1948.

Mark II in the old Gordon McKay Building at Harvard University, January 1948.

The central portion of the front panel of Mark II, with relay cubicles visible behind it, January 1947.

Front view of Mark III, September 1951.

The main control panel for Mark III, September 1951.

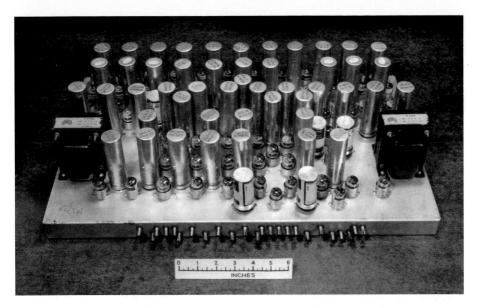

A typical chassis unit for Mark III, April 1948.

A magnetic drum storage unit for Mark III, September 1949.

The back of a row of Mark III's chassis, showing connectors, September 1951.

Aiken and two associates using the tape-preparation table for Mark III, September 1951.

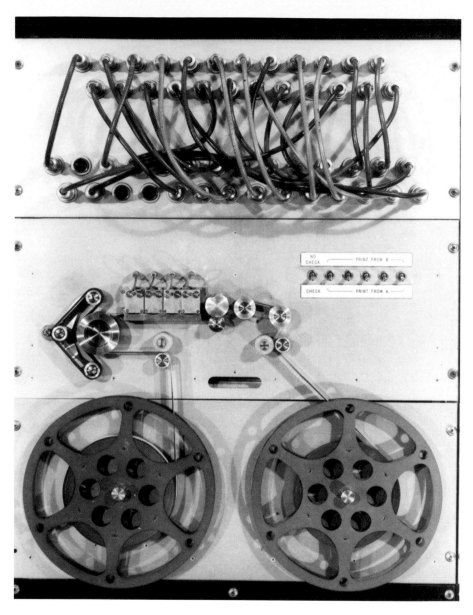

A printer tape mechanism for Mark III, September 1949.

Electric typewriters printing out results from Mark III, September 1951.

Assembling Mark IV, May 1951: rear view showing vacuum tubes.

Sequence and slow storage magnetic drum for Mark IV.

Front view of Mark IV during construction, March 1951.

Mark IV completed and operational in the Comp Lab, 1958.

Mark I (right), Mark IV (left), and UNIVAC (straight ahead) in the Comp Lab, October 1956.

Peter Calingaert, Howard Aiken, Frederick Brooks Jr., and William Wright on the occasion of Aiken's visit to the Department of Computer Science at the University of North Carolina, 1 March 1971. Courtesy University of North Carolina Photo Lab.

Aiken at Home, 1973

Henry Tropp

In the spring of 1971, soon after I had joined the AFIPS-Smithsonian project to record oral-history interviews with the major computer pioneers, I met with the advisory committee (composed of Isaac Auerbach, Walter Carlson, Cuthbert Hurd, Bernard Finn, Robert Multhauf, and Rudy Winnaker) to draw up a list of individuals who had been of importance in the development of computers and the art of computing. It was important to know the age and the health of every such individual in order to prioritize the interviews. Aiken was one of the giants, but neither his age nor his health targeted him for an immediate interview.

The advisory committee suggested that, before interviewing Aiken, I go to Cambridge to see what was left of Mark I, gain first-hand knowledge of its physical characteristics, its internal architecture, its components, its I/O, and so on, and talk to some of Aiken's associates. That visit included a number of fruitful discussions with Professor I. Bernard Cohen, who became one of my mentors, not only on Aiken but also on the early development of the computer. Bernard had recently been the primary historical consultant for the IBM Computer History Wall,[1] designed by the Office of Charles and Ray Eames and constructed at IBM's building at 590 Madison Avenue in New York, and he was then giving one of the first academic courses on the history of computing. He alerted me to the difficulty of getting Aiken to submit to an interview. Indeed, during the next year and a half, each time I called Aiken at his home in Fort Lauderdale he would refuse to be interviewed, with no hint that he might someday change his mind.

1. See *A Computer Perspective* (Harvard University Press, 1973 and 1983).

I decided that my only recourse was to get information about Aiken by interviewing those who had been associated with him. This list came to include Dick Bloch (the chief programmer for Mark I), Bob Campbell (Aiken's deputy during the completion of Mark I, and the chief designer under Aiken for Mark II), Grace Hopper (whose career began during the war years when she was assigned to Mark I), Bob Burns (Aiken's chief of staff during his last years at Harvard), various students (including Tony Oettinger, Fred Brooks, and Ken Iverson), and Bernard Cohen. One sad experience was my interview with Jack Nash at Lockheed. Jack asked me to turn off the tape recorder so that he could reminisce freely about Aiken, whom he obviously admired and with whom he had maintained contact over many years. Jack was very busy that morning, about to take off on a scheduled trip to Cape Canaveral to view a launching. He promised to arrange a more extensive session, but he died during the trip.

Gradually an image of Aiken began to emerge from discussions with those who had known or still knew him. The comments ranged from total admiration, such as that of Grace Hopper, to less admiring descriptions, a number of which began with "Just let me tell you about that . . . ." Every comment was at one of these extremes or the other, with no middle ground. A picture came into focus of a man who was physically gigantic and who had a strong personality and mindset.

Eventually, after many exchanges of letters between Bernard Cohen and Aiken, and especially through the happy intervention of Cuthbert Hurd, Aiken agreed to an interview with me if I were to be accompanied by Bernard Cohen. Bernard was not only a former associate of Aiken's, and someone who would guarantee that the interview would be up the highest standards of historical scholarship; he also had stressed to Aiken that his enormously important place in history was not being fully recognized and that it was therefore of the greatest importance to set the historical record straight.

Bernard and I met in Fort Lauderdale the evening before the first interview to plan our strategy. On one point we were clearly in agreement that there would be no beating about the bush. Bernard insisted that Aiken would welcome our being "blunt and direct." He proved to be right. Another decision, which in retrospect may have been a case of bad judgment, was that in the first sessions we would not probe every subject in depth, but would try to get an overview of Aiken's career and his observations on computer developments. We didn't want him to get going on Watson and IBM to such an extent that we might lose perspective on the other main events in his career. As it turned out, this was a sensible approach. Aiken made it clear to us, when we began the interview, that this would be only the first of a set of conversations, and that we could pursue certain special topics in depth and detail at a later time. We knew that we had to convince Aiken of our competence and seriousness of purpose and to get as broad a picture as we could before probing in depth the topics for future interviews. Unfortunately, since Aiken died three weeks later, some significant topics were never explored in depth and some not at all.

Aiken greeted Bernard Cohen and me warmly, with a courtesy and a degree of hospitality that showed why so many men and women respected and even idolized him. I was in awe of the man, but any lingering fear disappeared when he put us at ease and answered every question we posed with great attentiveness and seriousness. Many of his most interesting remarks were made at lunch or at dinner, or during one of the long walks we took together, and were not taped.

Aiken was a complex human being. He expressed his opinions forcefully and vigorously, but he allowed for—and even enjoyed—strong disagreement, especially from someone he respected. Once his mind was set, there was no getting him to change. In 1947, when Ed Berkeley was organizing the Eastern Association of Computing Machinery (now the Association for Computing Machinery), two of those who most strongly opposed the idea of such an organization were Howard Aiken and John von Neumann. Both felt strongly that there was no need for it, and that it would not advance the profession. Aiken believed there were already enough journals and societies in electrical engineering and applied mathematics. I asked Aiken if, in retrospect, he didn't think he had been mistaken. "No," he replied emphatically, "and I never did join and I never will."[2]

On one of our walks, Aiken told me about his youthful career as an electrical power engineer. I remarked that some of the people I had interviewed had told me that to understand the operation of Mark I you had to think of it as a large power plant. Aiken laughed, said "No," and promised to clear that up for me later. He never did.

2. I have never been able to verify the story that someone (sometimes said to be Bob Campbell) sent in a membership form in Aiken's name, and that Aiken became listed as a charter member.

When Cohen I were with him and when I was alone with him, Aiken was a generous host at lunch and dinner. Each dinner was in a different restaurant and was a gourmet treat, each one chosen because of some specialty that appealed to Aiken's fine sense of taste and love of good wines. The conversations were frank and delightful, both at lunches with Aiken alone and at dinners with Aiken and his wife Mary. I would be remiss if I did not record that I was completely over-whelmed by Aiken's personality and presence. If he was able to cast a spell, I freely succumbed.

In one conversation, I discovered a reason why Aiken was sometimes considered backward or even obstructionist. He refused to accept some novelty merely because it was new, the current state of the art. In the late 1940s Aiken had had a controversy with Norbert Wiener on the issue of punched cards versus magnetic tape for the input of a computer. In discussing this episode, Aiken remarked to me:

Well, the statement that punched cards were obsolete was even more ridicu-lous than to say that they're obsolete now, because they aren't. For input you can use either tape or cards, but cards—being discrete—have the advantage that you can separate them from the deck, you can mail them away, you can put them in a string, and you can sort them and re-order them. You can't reorder a tape. You can, of course, take data off it, put it in a machine, and then rearrange it and make another tape. But you can't order or change tape per se. Cards also have the advantage that when they are in a drawer they are subject to random access as well as serial machine access. So, if you think it over a little bit, you'll find that because we've had bright shiny magnetic tape, we've also had unsolvable problems.

According to Aiken, Wiener—upset because everybody wasn't rushing to use magnetic tape[3]—argued that "people who didn't give up old-fashioned procedures and go to the new ones were stupid."

Aiken had the reputation at Harvard of being difficult to work with on a committee, because he had his own ideas and was primarily interested in what would advance his own new subject of computers. I found, however, in my interviews with various mathematicians and computer scientists who had known and worked with him, that he was regularly asked to serve on committees for government agencies, uni-versities, and companies who were either looking to build or contract for the construction of a high-speed computer. Among those who served with him on committees were John von Neumann, John

3. In 1973 punched cards were still in wide use.

Curtiss, George Stibitz, and Franz Alt. In my interview with Curtiss, I found him to be a sincere admirer of Aiken. When Curtiss left the National Bureau of Standards to join the faculty at the University of Miami, he arranged for Aiken to work with him, helping to design a new computer center. He recalled for me how Aiken had come up with a brilliant plan that would permit the configuration of the facility to be changed easily.

Aiken's critics may have based their judgments of him on the effects of his influence in Europe in the years immediately after World War II. A number of the interviews indicated that, in Europe at that time, Aiken's stature in the computer domain was unmatched by that of any other single individual. His views were taken seriously and literally. Isaac Auerbach told me that, when he was organizing the meeting sponsored by UNESCO that led to the formation of the International Federation of Information Processing, it was essential for the Europeans that Aiken appear in a key role. His prestige, it was thought, would help the organizers to gain widespread European support. Fortunately for the future of IFIP, Aiken agreed to be the honorary chairman and to give welcoming and closing addresses. Aiken's stature in Europe was due in part to his willingness to share all he knew with anyone who was seriously interested—that is, with all who shared his particular view of what was the right way to build computers.

Aiken's contacts with European scientists may have originated during the war years. One of the stories that circulated about Aiken had to do with a signal from the French underground that they were in a position to "borrow" a new piece of German technology for just one night. Aiken was said to have been chosen as the person to study the new technology and to report on its use to the Allies. Supposedly, he was to be dropped secretly at a spot near Paris, where he would assess the equipment and then escape via submarine. He was said to have arrived a little early. It is hard to imagine Aiken, disguised as a French peasant, standing on a bridge or an overpass and waiting for his contact, but that's the story. In the version I heard, Aiken saw someone on the other end of the bridge or overpass and then—I now tell the story in the first person, as Aiken was supposed to tell it, according to my informant—"I watched this guy get closer and closer. I made up my mind that when he got close enough I would grab him, throw him over the bridge and run like hell. Just as I was about to grab him, this guy suddenly spoke up. 'Commander Aiken,' he said, 'you sure are one hard guy to keep up with. I'm trying to protect you.'" When I recited this story to Aiken, he kept a straight face, but his eyes had a bright twinkle in them. When I asked him point blank whether this story was true or was just one of the apocryphal tales circulating about him, he just looked at me and said not a word. He would neither confirm nor deny the truth of this episode. I was intrigued enough by the story to ask Grace Hopper if she had ever heard anything that even faintly resembled these events. "No," she replied. "But then, the old man would often be gone for some periods of time, and he would never tell us where he had been."

At one point in an interview, Aiken rose, walked to a closet, and brought out a small decorated tea cloth. He spread it out for me to admire, but I had no idea why he was displaying it. Then Aiken told me the story behind it. The cloth had been woven by a close friend, the Dutch physicist and mathematician Willem Van Der Poel. Van Der Poel, Aiken remarked, had worked closely with British Intelligence during the German occupation of Holland during World War II. To his contemporaries Van Der Poel appeared to be a collaborator because of his close association with Nazi officers. One of Van Der Poel's hobbies was weaving. He would weave these attractive cloths, which the Germans would buy to send home as gifts for their families. Unknown to the Germans, Van Der Poel had worked out a digital code in which he could weave information about such matters as the number and kinds of warships in the local harbor and their placement. These special cloths were then passed from hand to hand by special couriers and made their way to England, where the information could be used in planning bombing attacks or other military operations. According to Aiken, Van Der Poel suffered greatly because his fellow countrymen considered him a collaborator. It was only after the war had ended that his true services were recognized. When the Queen returned to Holland, she publicly awarded him a medal, and his reputation was restored. The cloth Aiken showed me had been given to him by Van Der Poel during a visit to Harvard. Aiken recalled how Van Der Poel learned to program Mark I and "would code the digits so that they would represent musical notes and Mark I would be typing out hymns that he composed."

Toward the end of our last session, Aiken and I were sitting in his study. On a footstool near my chair I had placed a pack of cigarettes and an ashtray. We were talking about some aspects of the current computer scene when he reached out, picked up my pack of cigarettes, held it in the palm of his hand, and said, "Hank, in the near future you're going to be able to pack as much computing power as there is in some of the current machines into something not much bigger than this." This was in 1973, before general miniaturization had become the order of the day. "You've got to be out of your mind," I responded. Bernard Cohen later told me that one of Grace Hopper's favorite Aiken stories was about an earlier prediction. She said that one could never tell whether Aiken was being serious or facetious about the future state of computing. The event she had in mind occurred in the late 1940s or early 1950s. What Aiken had said was that future computers with as much power as Mark I would fit into a shoebox. He almost lived to see that prediction come true.

IV

In His Own Words

Aiken in His Own Words
selected and introduced by Gregory W. Welch

Aiken and the Moore School Lectures

The Moore School Lectures on "The Theory and Techniques for Design of Electronic Digital Computers" organized by John Mauchly, J. Presper Eckert, and the Moore School of Engineering at the University of Pennsylvania, were given from 8 July to 31 August 1946 to an enrolled group of would-be computer scientists and engineers. The 52 lectures were given by a panel of leaders in the design, use, and development of computers. Howard Aiken presented the thirteenth and fourteenth lectures of the series, respectively titled "The Classification of Problems With Respect to Their Affect on Machine Design" and "Division and Root Extraction and Logarithmic and Trigonometric Function Values." The summary that follows is drawn exclusively from the lecture notes compiled by one of the students, Frank M. Verzuh.[1] The phrases quoted here are Verzuh's paraphrases of Aiken's words.

For the most part, Aiken's discussion during these two lectures focused on the experience of operating the IBM ASCC, or Harvard Mark I, and using it for scientific calculation. The technical aspects of his lectures (such as the operational components of the machine and the computational techniques for deriving the values of different functions), which make up the beginning of the first lecture and most of

1. Frank M. Verzuh, Personal Lecture Notes on the Theory and Techniques for Design of Electronic Digital Computers, unpublished typescript, 2 September 1946 (CBI 51, Charles Babbage Institute, University of Minnesota, Minneapolis). See Howard H. Aiken, "The Automatic Sequence Controlled Calculator" (lecture 13, 16 July 1946) and "Electro-Mechanical Tables of the Elementary Functions," in *The Moore School Lectures*, ed. M. Campbell-Kelly and M. Williams (MIT Press, 1985).

the second, were drawn directly from the Manual of Operation,[2] so we shall pass over them here. Instead we shall focus on what Aiken considered to be the strengths and weaknesses of Mark I's design, since his appraisal reveals much about his philosophy of computer design and operation.

The advantages Aiken attributed to the design of the ASCC/Mark I reflect his focus on creating machines to ease programming and to minimize the probabilities of errors arising in either coding or calculating. He cited the serial operation of Mark I as a distinct advantage of the machine. He also praised punched paper tape for program input because it permitted a program to be prepared "off-line," while Mark I was running a different problem, and reduced the time required to set up a problem on the machine. We may note that, by contrast, the task of re-wiring the ENIAC to tackle a new problem was laborious and time-consuming and required the machine to be inoperative. By including hardware in the ASCC/Mark I that automatically calculated the values of elementary transcendental functions (sine, $\log_{10}x$, 10^x), the amount of coding needed to express a problem was reduced with a resultant easing of the effort of programming. Aiken pointed to the flexibility offered by Mark I's storage registers and its ability to interpolate the value of functions from punched paper tapes: advantages, he testified, that had proven their worth in practice. Lastly, from an operational perspective, automatic checking procedures (using difference equations, identities, reproduction of results, and preassigned tolerances) contributed to the accuracy and reliability of the machine.

Aiken spoke candidly of Mark I's shortcomings. His comments indicate his determination to make the most efficient use of resources, both mechanical and human. He regretted the use of punched cards, in addition to punched tapes, because they could get out of order, thereby introducing errors. Mark I's constants registers (consisting of numerous rotary switches) proved cumbersome and unreliable; not only would the machine not function while all the constant values were being set, but setting them was a time consuming and laborious process that almost always resulted in a few incorrect values making their way into the problem. Aiken recommended eliminating the constants registers altogether. Another element of the machine that proved superfluous was the dividing unit. "No one should ever again build a

2. On the Manual of Operation, see Robert Campbell's chapter on Mark I.

dividing machine. There is one on Mark I, but that was a mistake."
He went on to describe how the iterative Newton-Raphson method
could be used to find the reciprocal of a number using only addition
and multiplication and how performing division by this series of calculations produced results faster than the special-purpose dividing
mechanism. Aiken also set his sights higher than the 93% efficiency
achieved by Mark I. Though the machine had run 5 weeks without
failure, Aiken wanted to increase its reliability. He also reported that
the original relays in Mark I were "poorly constructed." In order to
improve the machine's usefulness, Aiken suggested increasing the
number of "sequence devices" to four, and adding a "better automatic
check counter."

In response to questions, Aiken commented upon the design of
Mark II, at that time, moving rapidly from the drawing board to
construction. He placed "considerable emphasis" on the notion that
"the basic philosophy of computers should be to design machines to
[do] LOGIC. A LOGICIAN should cooperate in the machine design."

*This last point surfaces time and again in Aiken's lectures. Precisely how Aiken's
use of the term "logic" should be interpreted in the context of current computer
design is difficult to determine. It is clear, however, that he had in mind more
than the basic architecture of the processing unit.*

Harvard Symposia, 1947 and 1949

*In 1947 and 1949, Aiken's Harvard Computation Laboratory held two large
symposia, sponsored by the Navy, on "Large-Scale Digital Calculating Machinery."[3] These events were attended by scientists and engineers in the young field
of computing from around the world. Aiken's opening statements express his
views on the evolution of computing and the problems he judged to be of pressing
importance.*

3. The proceedings of these two symposia were published in the *Annals of the
Harvard Computation Laboratory* under the following titles: *Proceedings of a
Symposium on Large-Scale Digital Calculating Machinery, jointly sponsored by The
Navy Department Bureau of Ordnance and Harvard University at the Computation
Laboratory 7–10 January 1947* (Harvard University Press, 1948; MIT Press,
1985); *Proceedings of a Second Symposium on Large-Scale Digital Calculating
Machinery, jointly sponsored by The Navy Department Bureau of Ordnance and
Harvard University at the Computation Laboratory 13–16 September 1949* (Harvard
University Press, 1951).

In his opening remarks at the 1947 symposium, he stressed that the "number of people interested in this development" had increased so rapidly that "many of us have made discoveries only to have them superseded by later discoveries before the first could find application." Indeed, "the rate at which new discoveries have appeared has been such that none of us has been able to keep abreast of the situation as a whole." Aiken continued:

Throughout the past three years there has been inadequate publications of results, and inadequate transmittal of results from one group working in the field to another. Consequently, we have often found that we were beginning researches which were nearing completion in other laboratories. Those are precisely the reasons why this symposium seems so necessary at this time.

At the close of the symposium, Aiken identified the greatest challenge facing the new field:

The number of young men available to carry on the work in this field is far too small. It has become increasingly clear that we must start a training program. We must remember that our universities are primarily institutions for the building of men and not for the building of machines, and we must offer course of instruction in this field. I feel that one of the most important contributions the Staff of this Laboratory can now make is, with the coming of the next school year, to offer courses of instruction in applied mathematics, with a strong flavor of computing machinery. I sincerely hope that, with the permission of the Faculty of this University, we shall find and develop methods of furthering this purpose.

Two years later, at the opening of the 1949 symposium, Aiken once again stressed the need to meet the growing demand for computing specialists and outlined the steps he and his Harvard colleagues were taking to address this challenge:

I have often remarked that if all the computing machines under construction were to be completed, there would not be staff enough to operate them. Instruction in computing machinery represents one of the aggravated aspects of a generally recognized problem in technical education. We feel that the further development of the mathematical methods and the extended use of computing machines in the various fields represented by speakers here are those points at which levers should be placed to make the greatest possible advance in computer research. Only by completing computing machines and then operating them can the operating experience and experimental results be obtained that are so essential as a point of departing in passing from one design to another. Therefore, at our laboratory we have decided not to undertake the construction of any more large-scale computing machines with the exception of one, which we hope to build for our own use and keep at Harvard.

There is an ever-increasing number of industries interested in constructing computing machines outside the universities. In applying computing machinery to new and different fields, many proposals have been made, ranging all the way from devices for an automatic continuous audit, an automatic continuous inventory, down through an automatically operated insurance office, public-utility billing department, department-store accounting system, to more specific and less general accounting-machine components. Other proposals have included airline ticket-inventory systems, similar devices for railroad reservations, and automatic railroad ticket-vending machines. On the technical side, machines have been proposed involving automatic computers in connection with air-traffic control, airport control and almost every other manufacturing operation up to and including the automatic factory. But until our universities are able to offer well-rounded programs in numerical methods and the application of computing machinery to prepare men to operate these machines, the success of many of the proposed industrial programs will not be realized.

Large-Scale Digital Computers: Their Use and Limitations (speech before American Management Association Office Management Conference, 16–17 October 1952, New York)[4]

Although Aiken was a missionary for the cause of computing, he tempered his enthusiasm for the technology with heavy doses of realism, arising no doubt from

4. The photocopy of the transcript of this speech in the Harvard University Archives includes some later handwritten edits. Some of the edits clearly correct errors in the transcription; others make stylistic changes. The selections quoted here are from the original version.

Aiken's description of the "inventor" exhibited his disdain for those who do not "wish to disseminate the knowledge" they gain. "The scientific approach, on the other hand," he continued, "is more concerned with [discovering] why machines work than whether or not they work at all." "From this approach," he admitted, "we do not necessarily get marketable products immediately, but we do get information, science, and the lore on which to base developments in the future." Aiken then resumed his narrative.

There was, then, an end to this dark age around 1940, when the development of computing machinery and computing devices came back into the field of scientific investigation and a great many new developments made their appearance. As a result of these new machines, we find that . . . great hopes were encouraged on the part of those who wanted to use these machines in commercial and industrial operations. Inevitably, the popular picture of the large-scale computer as described in the press was far from the truth.

We at Harvard have recently completed a calculator which we call "Mark IV," This machine is a completely general-purpose, scientific computer. To the best of my knowledge, it would be utterly worthless in any commercial organization as far as accounting and bookkeeping are concerned. However, it illustrates the sort of thing which has been developed and what it might do in a business office if it were assembled in another kind of package, with different kinds of logic,[7] and operated in a totally different way.

Aiken then launched into a spirited discussion of computer technology and the specific technologies used in Mark IV. He described how vacuum tubes would be replaced by "new circular devices [magnetic cores] which take up less space." Storage registers may be built with these devices to retain "numbers (and the instructions as to what shall be done with those numbers)." A magnetic drum served "as the main storage unit for the vast quantity of information stored within Mark IV." Aiken explained:

We have already reached a point with our large computers where our chief problem is getting information into them and back out at a rate sufficiently fast to keep the computer busy. In short, we must have information going in on some physical medium, and we must get it

7. Here again we see Aiken using the word "logic" in reference to a concept along the lines of system design and organization.

back out on some physical medium. To manipulate this medium at something approaching electronic speed . . . is a serious problem.[8]

Aiken then turned to the value of Mark IV:

Anybody who wants a mathematical table and calculates it by hand is now wasting his time. In dealing with problems of airplane design, atomic energy, and radar antenna patterns—indeed a whole host of problems in which applied mathematics can assist by eliminating part of the experimentation leading toward optimum design—the large-scale computer pays off. There is no question that, where this type of problem exists, these machines will be of permanent value.

Aiken explained how Mark IV had proved useful to an engineer from the Bell Aircraft Company. Calculations that would have taken two years to solve and check by traditional hand and mechanical means were completed by Mark IV in two days. "This," Aiken exclaimed, "is not a situation in which the computing machine pays off. It is a situation in which it makes possible that which would not be attempted otherwise under any circumstances." Having attested to the computer's value to science, Aiken turned to its implications for "problems of accounting, industrial control, and so on":

It is indeed true that the arithmetical operations which this machine performs are those that are involved in accounting. It is indeed true that the way in which information is stored is exactly the type of thing which would be useful in accounting. But the fact remains that, while the existence of the large-scale, general purpose scientific computer may be regarded as experimental evidence in favor of the fact that it is possible to build an automatic office, it is not possible to take that machine as the standard and use it for that purpose. Rather, the techniques that have gone into the construction of this and other machines must be used in recast[ing] designs leading to operations of an entirely different character so far as the logics of the situation are concerned.

Altogether too little work has been done in this direction. In the past 10 or 12 years, we have reached a point where hundreds of agencies are interested in automatic computation, both here and abroad. Every European country is building computing machines. But the greatest

8. Magnetic tape served this purpose on Mark IV.

part of that effort is directed toward the development of general-purpose, scientific computers.

Aiken did, however, cite one counterexample: the construction of a large auto-mated mail sorter in Belgium that fundamentally changed the way mail was directed to its destination and permitted four operators to do the job of seventy mail clerks:

This mail sorter is big, clumsy and crude though it may represent the spirit in which the researches of the computing machine designer have been used. [N]ot in the sense that old equipment is to be improved in the light of new technology, but in the sense that the new technology is to strike out and to do a job in an utterly different way. . . . There are, then, two courses which are open to us. The electronic computing techniques can be used to improve our old devices and speed them up insofar as one can accomplish this purpose. On the other hand, we can start out anew on a more powerful basis to solve our problems in a much bolder way which will, in the long run, bring more important results.

To illustrate how an entirely new approach was needed when bringing comput-ing technologies to bear on classic problems, Aiken cited the absurdity in his eyes of building electronic replacements for the mechanical desk calculators widely used at the time:

Why pay another dollar for an adding machine that goes a thousand times faster when we are limited entirely by how fast we can [enter] the factors or the terms and how fast we can get out the results? I believe that the proper course lies in seeking utterly new ways of solving old problems, rather than attempting to improve well-known apparatus. . . . The big problems in developing new equipment are not the technical problems, but the problems of logics, which are interpreted in terms of what we want done and how we let our old procedures be changed in order that the new technologies can accom-plish the job. . . . If we are to give the machine designer the utmost opportunity to produce apparatus which will help us do a particular job, very often we are going to have to do that job in a different way.

In this context, Aiken identified the difficulties resulting from a "lack of indi-viduals who understand the situation as a whole." "The computer designer," he continued, "knows nothing about the insurance business, for example, and

*the insurance executive knows little about the detail of electronic procedures."
As a solution to this dilemma Aiken proposed "university courses in electronic
equipment for businessmen and training of young scientists for specific business
jobs."*

*Aiken also voiced his opinion of rapid obsolescence, which had long has
plagued those responsible for the implementation of computing technology:*

. . . There is a feeling on the part of many people that we should not
go ahead with the development of new equipment because there are
new technologies right around the corner. For example, transistors
may change computers completely. . . . Let me say emphatically that
this point of view will never win. Just as surely as we wait until the
transistor is available and perfected, there will be something else which
is just around the corner, and we shall go on waiting indefinitely before
we start our course of development.

*Aiken also debunked the popular perception of computers as "giant brains or
thinking machines." While he admitted that computers, and indeed many
machines, seemed to accomplish tasks that would require "thought" on the part
of human beings, Aiken's considered judgment was that "if we waste the verb
'to think' on operations of this kind, we have nothing left to describe those
processes that go on in the mind of man as he creates new ideas that previously
existed only beyond the totality of human knowledge." Rather than as "thinking
machines," Aiken preferred to consider computing machines as "useful tools—
greater and more powerful machines to serve the public and work for the benefit
of mankind."*

*Aiken, however, issued a word of warning to those who hoped that, by
surmounting the hurdles he mentioned, the computer would liberate mankind
from toil. "In the long run," he declared, "there is no such thing as a labor-
saving invention." For example:*

The automobile made transportation so easy we spent more time
traveling than ever before. Similarly electronic equipment will not do
all your work for you; it will simply enable you to do more work than
ever before. Use of this technology will solve many problems and also
will introduce a host of others you never met before.

Aiken on Mark IV:

Modern computers are made in large part from vacuum tubes. It is
my firm belief that the value of computers rests on the elimination of

vacuum tubes and the use of new circular devices [magnetic cores] which take less space. It is possible to store numbers in these devices, use the numbers over and over, reset the devices, supply them with new numbers, and so on. Storage registers of this kind, with numbers (and instructions as to what shall be done with those numbers), are in turn stored in a magnetic drum. On the surface of that magnetic drum these quantities are recorded and read by pole pieces. Such a drum is the main storage unit for the vast quantity of information stored within Mark IV.

We have already reached a point with our large computers where our chief problem is getting information into them and back out at a rate sufficiently fast to keep the computer busy. In short, we must have information going in on some physical medium, and we must get it back out on some physical medium. To manipulate this medium at something approaching electronic speed, so that the computing circuit will never load [overload?], is a serious problem.

The medium which we use is magnetic tape—paper tape coated with black oxide lining. By means of a device consisting of a central capstan to pull the tape, and over it a semi-stationary brass ring, it becomes possible to write or read 16 decimal digits in duplicate for checking purposes in about eight thousandths of a second. This device, developed further, supplies us with the means to get the information in and out of Mark IV at a speed which nearly approximates the speed of the machine.

Another step in the elimination of vacuum tubes is the use of small rectifiers—little discs that occur over one and three-quarters of the rotor. These are very cheap and small. When imbedded in a bakelite plate and joined together with silver conducting ink applied with a ruling pen, they form devices capable of multiplying decimal digits, adding decimal digits, and controlling a computing machine. These devices are covered and shielded from stray electrical effects and are plugged into the computing machine so that they may be rapidly changed when they get into trouble and lead into errors. The tubes also are plugged into the machine so that repairs may be made on the spot.

Mark IV has a main control panel and power supply which furnishes the 30 kilowatts necessary to its operation. In this main control panel are buttons by which the operator starts and stops the calculator and otherwise controls the machine. There are four typewriters that record the output results automatically—all data being set up in the form which one would expect of a type-setter, complete with page numbers, column headings, and so on, in preparation for publication by offset lithography.

The State of the Art of Electronic Computers (talk at Business Conference at Harvard Graduate School of Business Administration, June 1954)[9]

This talk, in which Aiken once again examined the applicability of computers to business, offers further insights into his view of computers as complex systems configured and adapted specifically to the task at hand.

Aiken began with a humorous dialog illustrating the different ways of viewing computers. The speakers were Groucho Marx, who conducted the interview, Dr. Harry Huskey, a computer designer for the Bureau of Standards, and "a fellow by the name of Sol, the most distinguished junk dealer in Los Angeles."

Groucho: Mr. Huskey, I understand that you have just completed the construction of a great machine that almost thinks.

Harry Huskey: Well, yes, that is right. I have completed a big computer, and it is called "the SWAC."

Groucho: Well, now, what do you think that machine is worth?

Harry Huskey: Well, I don't mind telling you that it set the company $400,000 to build the machine.

Groucho: Sol, what do you think the machine is worth?

Sol: I guess I can't give you an opinion, because I don't know anything about it. I've got to know a little more about it, before I can answer a question like that.

Harry Huskey: Well, it's a two-megacycle repetition rate machine, with forty binary digits and register . . . .

Sol: But I want to know the essential facts about it. How much copper is in the machine, and other materials?

It seems to me that this story has relevance for those who are interested in applying the machine in business. Each of those two men gave his way of judging the possible worth of the computing machine. I suspect that somewhere between these limits lies the opinion of every [member of the audience].

Before speculating on the future of computers, Aiken, as was his custom, looked to the past for insights:

---

9. Excerpts taken from a transcript in the Harvard University Archives.

I do not exaggerate when I say that everyone who was in any way connected with the original development of large-scale computers had in mind only one thing—namely, the construction of machines for the solution of *scientific* problems. . . . No one was more surprised than I when I found out that these machines were ultimately going to be instruments which could be used for control in business.

Aiken attested to the successful implementation of scientific computers, particularly in the field of compiling tables, such that "manual labor in this direction can be abandoned":

So far so good. But how does this lead us to the point where we begin to wonder whether the computer can be used in business? . . . Scientific computation makes use of the four fundamental rules of arithmetic. . . . Accounting uses the same rules. . . . In scientific computations, you have to store up certain numbers while you do something arithmetical with certain other numbers, then you recombine them and so on. Accounting follows exactly the same procedure. Therefore, since we already have a computer for solving scientific problems, why could we not apply it to the solution of accounting problems as well?

From an arithmetic point of view, fundamentally, there is absolutely no reason why this could not be done.

However, this observation led to much discussion and confusion, Aiken claimed. One week you could hear a distinguished speaker extolling the computer as a practical and available tool for business, and the next week you could listen to an equally distinguished speaker who "throws cold water on the whole thing." The difficulty arises at the point where one attempts to take the philosophical applicability of scientific computers and convert it into a real business machine:

. . . when you start examining the property of these machines—the peripheral equipment, the input equipment, the output equipment, and so on—you will find that there can be vast differences in accounting and scientific applications. . . . From the standpoint of linear programming, statistical analysis, and any kind of computations of a mathematical character that are useful in helping to operate a business, there is no question but that electronic computing machinery is applicable [to business]. Moreover, there is no question but that the machines already produced are adequate to undertake these tasks.

However, the issue was not so clear when one examined the problems faced by someone who wished "to turn out 50,000 gas bills per day." In Aiken's opinion, at the time, anyone who looked to commercially available computers as a means of economically automating this process could be "led to some wishful thinking." Aiken sought to dispel an "almost unbounded" enthusiasm for computers by educating his audience about the tradeoffs and realities that must be faced in applying computers.

The crux of Aiken's argument was the issue of economic feasibility. At a time when a computer cost hundreds of thousands of dollars, Aiken advised that before purchasing such a machine a company should consider whether the purposes in mind "are best handled by large-scale computers or by lesser devices of a single-purpose character." In addition, he advised, businesses must recognize that the "tools which one uses for carrying out any piece of work affect, at least in part, the nature of the product." That is, "when machines are applied to carry out operations, in all probability there will [have to] be changes in the operations [of the organization]." Aiken cited as an example the changes and standards banks would have to adopt if they hoped to mechanize their operations:

Machinery cannot be expected to carry out in the same way the operations that are presently done in offices by manual means. . . . There must be give-and-take on the part of those conducting the operations of the system, if the machines are to go into operation. And unless you are willing to study the machinery itself enough to know the concessions you will have to make, you will not reap the rewards available to you.

Similarly, Aiken sought to dispel a prevalent misconception that computers could actually "solve" problems:

It is true that the computing machine, because of its great power and speed, can accomplish things that a person would not live long enough to do. . . . But [in]sofar as *how* operations are to be done, a person has to be completely able to know and plan this for the machine . . . No computing machine can do for a person anything more that he cannot think through himself.

Once more, Aiken concluded by refuting the idea that computers could think:

If we can get further and further away from this notion of thinking machines and realize that we have to do the thinking, that we have to understand the problem and know how to program the computing machine . . . then the computing machine will be a highly useful tool.

He acknowledged that he may have dashed the hopes of some in the audience:

It may sound as though I am debunking the machines. [Indeed] something of that feeling is what I wish to convey. But it isn't the machine itself that I want to debunk. . . . I have the greatest hopes for computing machines.

The Future of Electronic Computing Machinery (1956)

In 1956, an article entitled "The Future of Automatic Computing Machinery" appeared in the German publication Elektronische Rechenmaschinen und Informationsverarbeitung, *issued by the Technische Hochschule at Darmstadt, where Aiken had given this lecture at the invitation of his friend and colleague Alwyn Walther. The editor's note indicates that it is a transcription from a tape recording and that it had not been revised by the author. A copy is in the Harvard University Archives, along with a transcript of a talk given in Sweden that is almost word-for-word equivalent to the one given in Germany. Either Aiken was reading from the same prepared text on both occasions or perhaps he had the transcript of one of the talks in hand when he gave the second one.[10]*

Though the talk echoes many of the themes of the previous two lectures, here Aiken did not focus on such historical figures as Pascal, Leibnitz, and Babbage; rather, he offered a historical perspective on the 10-year development of the computing field. In this talk, Aiken almost paints himself as a historical figure.

Aiken prefaced his discussion of the development of the field by recalling the opinion of his teacher at the University of Wisconsin, Professor Edward Bennett, that the "development of a new body of knowledge always passes through four stages": observation, classification, deduction, and mystification. The fourth phase he defined as when "the demonstrations of the masters were accepted as facts without question and a super-abundance of words stifled thought." "I can personally attest," he continued, "that when a few men began the modern effort in large-scale computers a decade or so ago their work began

10. The Swedish transcription, which bears the title "The Statesmanship of Computers," is dated 7 November 1955. At the bottom of each page there is the mark "1956:110 A." The selections printed here come from the published Darmstadt version.

with observation [the first stage]. . . . If there is any question in your mind that we are in the fourth stage to-day, I ask you only to make a list of the names we have given our computers."

A decade or so ago, when we began [work] with large computers, there were but a few men working in this field. Indeed, it was possible for each and every one of them to know all of the others. I don't know . . . how many people are engaged in this effort to-day . . . but the number of people involved . . . is certainly of the order of 10,000 throughout the world. In fact, we have now progressed from the point where we first could know all of the people, to the point where now we cannot remember all the names of the machines that have been built.

Aiken estimated that the worldwide investment in computing by 1956 had reached "somewhere in the vicinity of 500 million dollars." "What," he asked, "have we achieved with these 10,000 people and what have we bought with [this] fabulous sum of money?"

To make a survey of the situation let us go back and note that when our modern effort in large computers was begun, those who were concerned with the problem had in mind only a single objective: the mechanization of numerical analysis. We wanted neither more nor less than this. This means, of course, that we are interested in machines that could manipulate relatively large linear systems, automatically integrate differential equations and systems of equations, tabulate the values of different functions, and so on. . . . In short we were primarily and solely interested in solution of the mathematical problems which arise in physics and engineering and technics in general.

This is, of course, an unending task, because we are constantly presented with new problems arising in science, technology, economics, and so on . . . [and] theoreticians constantly provide us with new methods for the solution of these problems. More important still, the situations which they expect us to deal with [with our] machines are of ever-increasing complexity. While they were satisfied with 30 or 40 linear equations ten years ago, we now find people talking about linear systems of the order of 400 or 500 for example.

Despite all this and with modest exaggeration I feel that we are justified when we say that the original objectives, which we set ourselves, have in a very large part been obtained, because we have produced machines which could do what we set out to do—and a lot more. Not only that, but these machines have been produced and have become so common that they have been reduced to articles of commerce. This means it is no longer necessary each time we require a machine to begin in basic research, develop the components, develop the design, build the machine, as most of us originally had to do.

In his assessment of the accomplishments of the field, Aiken also touched upon the development of switching theory and the invention of a "whole host of new switches" including "the rectifier, the transistor, and the magnetic core," and "new storage devices, of which the magnetic core and the ferro-electric matrix represent those which are of greatest important to-day."

Aiken then turned to establishing an agenda for future efforts in the field: " . . . one of the prime responsibilities which rest upon us . . . is the continuing search for new switches, new storage devices of great power, lower cost, in order that our art may make further progress." Indeed, Aiken "felt confident that the next real advances . . . will be made in part by the metallurgist . . . [and] the mechanical engineer in the development of tools and processes for [manufacturing computers]." Such developments would, in turn, lead to progress in the "so-called new applications" of data processing and process control. Reminding his audience that the concept of "the elimination of physical drudgery" resting in "the development of automatic machines" dates back to Aristotle and motivated Pascal and Jacquard alike, he pointed out that "we are presented with new opportunities which transcend the imagination of those who went before us in attacking a number of problems by automating tiresome chores. Yet:

Why is it with these opportunities and with this new equipment . . . that we make progress so slowly? Why is it with our new logics, new switches, our new techniques, and storage devices, that we talk about the future of automatic control and the future of data processing?

The first answer is that [for the most part] all our technicians, all our computer engineers [for] the last decade or so have been preoccupied with scientific computation. [While] I would in [no] way detract from that effort . . . [since] there is no question that, as time goes on, we will need larger and more powerful machines for scientific work, . . . I would like to see us renew our efforts in other applications and to recognize the importance to society in general of data processing and automatic control.

The second reason cited by Aiken for the sluggish progress in the development of these applications is more difficult to grasp and has indeed been cited as evidence that he did not fully understand the nature of the computer:

In part we have been trapped by an ambition which is common to the human race, namely, that we try to solve all our problems at one fell swoop. We have perfected the general-purpose computer for scientific computations. Now that same device with little alteration we attempt to employ in a great variety of situations for which it was never intended in the first place. Under these circumstances, if it should ever turn out that the basic logics of a machine designed for the numerical solution of differential equations coincide with the basic logics of a machine intended to make bills for a department store, I would regard this as the most amazing coincidence that I have ever encountered.

Today, when we use the term "general-purpose computer" we have in mind a machine that can undertake any task that can be described by a series of logical instructions. As Alan Turing proved, in principle, any computer can perform any "computable" task. In this light, Aiken's comment seems to betray an ignorance of the true "general-purpose" nature of the computer. However, in practice, whether a given computer can efficiently perform a given task is an entirely different matter.

To interpret precisely what Aiken meant, we must understand his use of the term "basic logics." We gain a further glimpse into his intention some paragraphs later, when, comparing the machine requirements for scientific computation with those for data processing, he says:

Logics, the amount of logical operations, by this I mean binary decisions,[11] in scientific computations is so small, that a flip-flop or some similar device is quite sufficient for our purposes. Seldom, if ever, have we built logical decision equipment as such in scientific computing machines. On the other hand, the number of binary decisions to be made in most data processing operations are far greater in number than the arithmetic operations, and hence, the inclusion of circuitry for logical operations is essential in data processing machinery.

Aiken also noted that high capacity input and output was much more critical for data processing applications than it was for large internal storage and

11. What today we would call conditional branches.

circuitry to facilitate rapid sorting of data. Scientific computation benefited from machines that could rapidly perform arithmetic. "But in data processing, most of the operations on large computers that I have seen require the arithmetic unit to operate not more than 4 or 5% of the time. Hence, here is an opportunity to cut the amount of equipment involved, simplify the design, get rid of the cost, the heat generated and the refrigeration necessary to take the heat away." These comments clearly reveal that Aiken is not espousing his views in the abstract realm of computation theory, but in the real world of system design and practical engineering.

When Aiken began his career, computer designs were not mass-produced; they were created for a specific institution and a specific type of application, which designers, for economic reasons, had to take into account as they built the system essentially from the ground up. Though in the course of his lectures we have seen him use the word in several different senses, we are drawn to the conclusion that Aiken has stretched his meaning of "logics" to embrace the larger relation- ship between the structure of the computer system as a whole (from processor circuitry to input and output peripherals), the problems to which the system is to be applied, and the overall organizational context in which it is to be used. Given this meaning, Aiken's claim makes perfect sense.[12] In today's world of mail-order computers, Aiken's "logics" are the stock in trade of system analysts who select from a host of specialized "general-purpose" computing devices to craft a system that best meets the needs of their clients.

That Aiken actually did appreciate the principle of the general-purpose computer is clear from the fact that, having contrasted scientific computer and business data processors, he turned to the "one fundamental similarity between these devices . . . and that is that the general purpose of the digital computer is in many ways the model of an automatic process." He proved his point by analyzing a screw-cutting machine and the telephone system in terms of the basic components of a computer: input, storage, arithmetic unit, sequence control, and output. Thus, in the automation of any system, Aiken claimed, "our problem is to describe the outputs that are required, to describe the necessary inputs to produce these outputs, to write out the logical equations that connect the inputs with the outputs, then with the logical expressions which connect these, to discover in their transformation the design of the machine which is necessary to implement the task in mind."

12. Before the introduction of the IBM System/360 (1963), IBM and many other computer companies had two separate lines of computers for scientific and business applications. Even the processor and peripherals chosen for the 360 family were generally dictated by the intended application.

Aiken then returned to why progress in data processing had lagged behind its promise. The greatest hurdle, he argued, was that, in contrast to mathematics, there existed no "unique language" in which to describe the operations that must be automated in business and control environments. In addition, automating an accounting office or an oil refinery was "[by its] very nature interdisciplinary in character." In order for progress to be made in this realm, Aiken asserted, "the institutes of technology and schools of engineering in our universities" must be willing to "broaden the sphere of our own activities . . . to cut horizontally across the vertical divisions between the several disciplines and activities of the race as a whole."

What Is a Computer?

The manuscript[13] of this discussion is not dated, but it bears the handwritten notation "Wayne University." It would appear to be the closing remarks of a conference on computing. In answer to the question posed by the title, Aiken basically answers "It all depends on how you look at it." He offers an anecdote as illustration:

Some years ago I was visited by a young man from a very large corporation who had come with plans for a computing machine he intended to build for his company. He wanted my advice and approval for his plans. Most of the day had gone in fruitless discussion when he finally said: "I think I see why we are not getting anywhere. You are interested in operating computing machines and I am interested in building them."

In part, according to Aiken, the many different perceptions of computers arose from the fact that a diversity of disciplines were involved in getting any use out of the machines. "As functional installations," he noted, computing centers "are showing themselves to be of broad versatility, cutting a wide swath across traditional lines of scientific and industrial activity."

Once more Aiken prescribed an interdisciplinary approach to education. "Such a breadth of scope" was "required because of the enormous diversity" of potential applications. Even "at [their] best," he argued, our educational institutions "will not be able to develop the required men as rapidly as they begin to be needed now in industry." We must "learn from each other how to meet these problems which are expanding so fast," Aiken exhorted his audience. "It will

13. In the Harvard University Archives.

contribute to our success if everyone working with computing machines comes to appreciate his own limitations. Broadening our sympathies in this manner, we shall have gone a long way in advancing the subject and to fill the interim requirements until the slow progress of higher education has adjusted itself to the new demand."

International Conference on Information Processing, 15–20 June 1959

The conference, held in Paris under the sponsorship of UNESCO, marked the first major step in the formation of the International Federation of Information Processing Societies (IFIPS, later IFIP).[14] It brought together more than 2000 computer scientists from around the globe to discuss developments in the field and its future. The organizers sought an august figure to preside over the proceedings—someone who would be known and respected by all the participants, especially the Europeans. Their natural choice was Howard Aiken.

Aiken's opening and closing remarks,[15] which bracketed this first large international congregation of computer professionals, suited his stature. He spoke as one of that small group who had given birth to the field, who had witnessed or participated in most of its significant advances, and who knew its strengths and its weaknesses.

In his closing remarks, Aiken graciously thanked those whose efforts had made the conference possible, including UNESCO, and then briefly summarized the activities of the conference: 59 papers were presented in 11 sessions, and 60 lectures were delivered in 12 symposia. To set a tone and a context for what was to come, he offered his own vision of the past, present, and future of computing. From its earliest phase, in which a handful of men like himself sought merely "to demonstrate the feasibility of that which we wished to undertake," the field moved quickly "to show that we could indeed produce the required results." Next came "the programme for the improvement of machine components, such as switches and storage devices," which in turn "brought us to present developments which are largely dependent upon solid state physics, and perhaps, in the near future upon electro-chemistry." Such developments "will continue so long as faster and faster machines are required, or so long as machines of small bulk and greater internal power can be conceived by men." With respect to theory, he said, efforts have resulted in "an improved theory of

14. For details see *A Quarter Century of IFIP*, ed. H. Zemanek (North-Holland, 1986), especially p. 44ff.

15. Edited transcripts of the opening and closing remarks are in the Harvard University Archives.

switching, which is largely replacing the empirical design techniques employed a dozen years ago in building computing machines" and "have laid the foundations of a theory of systems,[16] *which represents perhaps one of the most badly needed requirements of modern times."*

Among the more recent accomplishments of the field, Aiken cited "the application of the ideas of automatic machinery to a variety of other activities." From an economic perspective, commercial data processing was the most significant. Turning his eye toward the future, Aiken noted that close behind data processing would follow "the application of computing techniques to the control of machine tools and plant facilities, which [though it was not addressed at the conference] awaits the effort of the people here at this Conference." Other promising new applications for computers in his opinion included "the automatic translation of languages, applications of computer-like devices in medicine, the preparation of scientific abstracts, the composition of music, the production of concordances, to mention but a few."

Aiken rallied his audience with a vision of the important role computing would play in the world:

[At] the present time then, the information processing aspects of nearly every sequence of human activity begins to fall with the area of interest of those who will process that information by machine. Perhaps no group attending any congress, ever faced so many challenging problems so boldly and with so great a chance of success or with such great a potential influence on society as a whole. Accordingly, in closing these remarks, I cannot refrain from observing that with opportunity, with power and capability, comes also responsibility; namely, in our case, the responsibility for the wise application of the computer . . . in the public interest and for the benefit of society in general.

Closing Remarks, International Congress on Automation, Madrid, 1961

The last known speech of Aiken's Harvard career as a computer scientist was given at an international meeting of computer scientists held in Madrid and organized by José G. Santameses, who had spent some time working with Aiken at the Harvard Computation Laboratory. Aiken's closing remarks highlight some of the changes in the field since he helped give it birth. In summarizing the activities of the Congress, Aiken noted that its 340 registered participants

---

16. At last a term had been developed that captured the meaning Aiken had tried to convey by using "logic" for more than 10 years.

represented 19 different countries, 45 government agencies, 70 industrial organizations, and 30 universities, and that they had presented papers on six general topics: commercially available computer systems, data processing applications of computers, computer control of machine tools and manufacturing processes, the "study of machine components," switching theory, and "the economic and humanistic importance of automatic machinery in general."

In offering his personal perspective on these presentations, Aiken observed that, to judge from the papers on commercially available systems, "the design of computing machinery has begun to be stabilized. It is no longer the case that a machine may be obsolete by the time it has first been constructed. . . . Advances in computing machine construction, at the moment, proceed with the aid of development rather than basic research."

With respect to data processing, Aiken continued to express dissatisfaction with the automation of business operations. The speakers at the Congress had explored two separate approaches to bringing the benefits of computers to commerce: applying "large-scale computers, through proper programming," and "the construction of computer-like devices, using computer-type components, for the solution of specific data processing problems." Yet, despite the "vast economic importance" of this field and the many impatient customers yearning to reap cost savings through the use of computers and the "feverish activity" by suppliers eager to satisfy this demand, "the design of data processing systems falls under the heading of system engineering, and systems engineering is an area in which theory is not too well developed. We have not succinct mathematical means for describing business and other data processing systems, hence it is very difficult to deal with such systems, except as they may be described in words; and further, because of this very difficulty, it is too often hard to distinguish the virtues of one system over another."

Lest any of his listeners misconstrue his remarks, Aiken hastened to make his message explicitly clear: "These remarks should not be understood as a criticism of our speakers, but rather as an expression of a serious need in the computing machine field. These remarks have been made in the hope that they may help stimulate others to work in this area, developing better systems-logical ideas, and better techniques for the mathematical treatment of systems." Aiken's belief that it was necessary to issue such a caveat may serve to indicate his awareness that his insistence on this point was less than well received by his peers.

Like data processing, the application of computers to automatic control problems suffered, in Aiken's eyes, from the lack of a rigorous approach to system design and from its very interdisciplinary nature. Education must rise to the challenge of providing cross-disciplinary training.

At the sessions, most of the papers concerning research in computer compo-nents had dealt with "ferro-resonant techniques." Indeed, despite his earlier assessment of the stability of computer design, Aiken stressed that "it should never be forgotten by computer engineers that the discovery of new switches and of new storage devices represent two of the most important responsibilities which rest in their hands. Hence, it can only be hoped that work in this area will be accelerated, particularly making the fullest possible use of solid state physics techniques."

Some critics have wrongly accused Aiken of failing to embrace electronics. Indeed, what Aiken disapproved of was the dependence upon vacuum tubes, which in his opinion were inherently unreliable. However, he saw early the promise of solid-state devices (eventually transistors) for computers, espoused their acceptance, and introduced germanium crystal diodes in both Mark III and Mark IV.

Aiken believed he could add little to the many papers presented on the "theory of switching . . . other than that the importance of this discipline to the com-puting machine engineer could not be overstated. The mathematical basis for a theory of computing machines and for a theory of systems . . . may well rest in the theory of switching."

Finally, in commenting on the economic assessment of the technology offered by several of the speakers, and no doubt in response to a swelling public concern over the long-term implications of automation, Aiken observed:

History has made it extremely clear that the effect of automatic machines has been, and [will] continue to be, that we produce a greater output per worker; we make use of a greater installed horsepower per worker, and as a result of these, we provide an improved standard of living for the worker, and above all, we eliminate drudgery and boredom.

Appendixes

Specifications of Aiken's Four Machines
Robert V. D. Campbell and Peter Strong

AUTOMATIC SEQUENCE CONTROLLED CALCULATOR/MARK I FACT SHEET (original configuration of machine)

A. COMPLEMENT OF FUNCTIONAL EQUIPMENT

Internal

72 accumulating electro-mechanical Storage Registers

1 Multiply/Divide Unit

3 Built-in function units: $\log_{10}x$, $\exp_{10}x$, $\sin x$

3 Interpolator control units

Special features: choice counter, hi-accuracy computation

Input/Output

60 sets of dial switches for inputting numerical constants

2 punched card readers

3 functional data punched tape readers (with automatic tape positioning)

1 instructions punched tape reader

2 electric typewriters

1 serial card punch

Off-Line

1 Manual (keyboard) tape preparation unit (Serial card punch can also be used manually)

B. PRINCIPAL COMPONENTS

2204 electro-mechanical rotary decimal counters (accumulators)

3304 electro-mechanical relays (control)

225 cam-operated contactors (circuit breakers; electrical timing)

Standard punched card readers and punch—80 column cards

Custom punched tape equipment—24 rows of holes plus 2 sprocket holes

24 wire transfer bus for data words

C. DATA WORD FORMAT

24 decimal columns, utilized as 23 columns + sign

decimal point position is adjustable by plugboard

internal transfers are 24 decimal digits in parallel, serial within a decimal digit requires 4 rows of holes on function tape

D. INSTRUCTION WORD FORMAT

24 bits, partitioned into 3 8-bit fields

one instruction controls data word transfer and initiation of an operation

fields are assigned: out address, in address, operation code; address & operation codes are pure binary frequently either out address or in address implies the operation requires 1 row of holes on instruction punched tape

E. OPERATING TIMES

basic cycle time 0.3 sec., partitioned into 16 impulse times

add, subtract, transfer, read card—	0.3 sec.	1 cycle
multiply—	up to 6.0	20
divide—	up to 15.6	52
$\log_{10}x$—	up to 89.4	298
$\exp_{10}x$—	up to 65.4	218
$\sin x$—	up to 60.0	99
print one data word (max.)	6.9	23
punch 24 columns of card	3.0	10

Note: add, subtract, transfer, various other operations, can be carried out during multiply or divide.

F. SOURCE OF KEY COMPONENTS

All components were provided by IBM. Some were standard commercial units and some were components invented previously, but not yet used in commercial equipment. The tape readers & punches were custom designed for the project.

G. CHRONOLOGY

1937 Initial system concept

1939 Started design and development at IBM in Endicott, NY

1944 (Winter) moved to Research Laboratory of Physics, Harvard, after completion of assembly and testing at Endicott

1944 (August) dedicated; in operation as Bu Ships project

1946 Moved to new laboratory building

1959 Retired from use

H. ENHANCEMENTS TO ORIGINAL CONFIGURATION

Special registers for conditional and unconditional transfers of control; free standing subsequence unit

Electronic multiply unit

Additional storage registers

Interpolator tape unit adapted to read instruction tape

MARK II FACT SHEET

A. COMPLEMENT OF FUNCTIONAL EQUIPMENT

Note: On-line equipment comprises two complete machines, a "right" machine and a "left" machine, which can be operated either together or independently.

Internal—for each half machine (all operations are performed electro-mechanically)

48 electro-mechanical relay storage registers

1 addition unit

2 multiply units

6 built-in functional units: x^{-1}, $x^{-1/2}$, $\log_{10}x$, $\exp_{10}x$, $\cos x$, $\arctan x$

Special functional registers: transfer, cross, code interchange, check, start-stop . . .

Controls for interpolation process.

Input/Output—for each half machine

12 sets of dial switches for inputting numerical constants

2 numerical data punched paper tape readers—for sequential inputs

2 interpolator (functional) data punched paper high speed tape readers—with automatic tape positioning capabilities

2 numerical data tape-punching mechanisms

2 teleprinters

2 instruction punched paper tape high speed reading mechanisms

Off-Line—one set serving both half machines

1 Manual (keyboard) instruction tape preparation unit

1 Manual (keyboard) data tape preparation unit

B. PRINCIPAL COMPONENTS (for both half machines combined)

13,000 electro-mechanical relays for data storage and control, some of which were latching-type relays for data storage without power drain.

800 cam-operated contactors (circuit breakers; electrical timing)

46 wire transfer bus for data words

8 5-channel punched tape readers for data (5 rows of holes plus 1 row for sprocket holes)

4 tape punching mechanisms—same format

4 teleprinters

4 6-channel punched paper tape readers for instructions (6 rows of holes plus 1 row for sprocket holes)

C. DATA WORD FORMAT

Floating point data word:

"Mantissa": ten decimal digits, represented in binary coded decimal (bcd).

Exponent (power of ten): 4 binary digits, providing 0 through 15.

Two algebraic signs: "Mantissa" and exponent.

46 relays (one per bit) are required to represent word.

On numeric tape, one word requires 13 rows of holes[:]

1 index

1 both signs

10 ten decimal digits

1 exponent.

Internal data transfers are fully parallel, using 46 wires.

D. INSTRUCTION WORD FORMAT (external, octal; internal, binary coded octal)

Represented by 18 bits, partitioned into 3 fields.

Fields are: sign, out address *or* in address, operation.

Bits are assigned as follows:

Sign: 1 octal, 3 bits

Out address or in address: 3 octal, 9 bits

Operation code: 2 octal, 6 bits

On instruction tape, represented by 3 rows of holes, 6 per row.

(*Note:* programming was done in a highly structured format. For each instruction, either the out address or the in address was supplied by the machine.)

E. OPERATING TIMES

(*Note:* On either the right or left calculator, can perform 4 additions, 6 transfers and 2 multiplications each second.)

One pulse time is 1/60 sec., computing cycle time is 1 sec.

Execution of 1 data transfer takes 2 pulse times: first to set up control relays, second to effect transfer. Hence, basic instruction execute time is 1/30 sec.

Times for floating point operations:

Addition 0.2 sec.

Multiplication 0.7 sec.

Functions: division	(x^{-1}, y)	4.7 sec.
	$x^{-1/2}$	6.0
	$\log_{10}x$	5.2
	$\exp_{10}x$	6.7
	$\cos x$	7.5
	$\arctan x$	9.5

reading data words from sequential data tape—6 words per sec.

positioning & reading values from interpolation data tape ~8.0 sec

reading from instruction tape—30 instructions/sec. (30 rows of holes are read, 3 times per sec.)

punching an output data word into data tape—1.5 sec.

printing one data word—4.0 sec.

F. SOURCE OF KEY COMPONENTS

Relays—Autocall Company, custom designed to Comp. Lab. specifications.

Punched paper tape equipment and teleprinters—Western Union Telegraph Co., except High Speed tape readers, designed and built by Comp. Lab.

G. CHRONOLOGY

1944 (November) Initial system concept

1945 (February or March) Started design

1948 (February) Moved to Dahlgren Naval Proving Ground (now the Naval Weapons Laboratory) after assembly and test at Harvard

1948 (September) Started production operation

1955 Retired from use, some time after delivery of NORC computer to Dahlgren

MARK III FACT SHEET

A. COMPLEMENT OF FUNCTIONAL EQUIPMENT

Internal

200 data words (fast access) on magnetic drums

150 constant data words (ROM, fast access) on magnetic drums

10 constant data words (fast access) on magnetic drum, read in from manually set dial switches

special data registers in-fast storage

4000 data words (slow access) on magnetic drums

1 electronic adder

1 electronic multiplier

6 built-in functions: x^{-1}, $x^{-1/2}$, $\log_{10}x$, $\exp_{10}x$, $\cos x$, $\arctan x$

controls for interpolation process

4000 words of instructional storage on magnetic drum (capabilities for multiple precision arithmetic)

Input/Output

8 magnetic tape read/write units for data

1 magnetic tape read unit for instructions

10 sets of decimal switches for introducing constants

Off-Line

5 electric typewriters

5 magnetic tape data readers for printer output

1 manual (keyboard) data tape preparation unit

1 manual (keyboard) instruction tape preparation unit, with many built-in programming aids

B. PRINCIPAL COMPONENTS

8 horizontal magnetic drums for data (6900 rpm, 8 in. dia., 10 pulses, in., 27.6 kilohz.)

1 vertical magnetic drum for instructions (1725 rpm, 16 in. dia., 20 pulses/in., 27.6 kilohz.)

4500 vacuum tubes of 7 types: triodes, dual triodes, pentodes and twin diodes; germanium crystal diodes

Electro-mechanical relays

15 4-channel 5/8 in. mag. tape units for data tape: channel a—data bits, channel b—duplicate recording of data bits, channels c and d— timing.

Record 50 bits/inch in each channel, read/write at 0.8 kilohz.

1 4-channel 5/8 in. mag. tape reader for instructions. (Instruction tape is similar, but channels *a* and *b* re not duplicates.)

data transfer bus—4 wire

C. DATA WORD FORMAT

Seventeen decimal digits (17th is just used for sign).

Decimal point is adjustable.

Each decimal digit uses modified binary coded decimal (mbcd) code: columns are weighted 2*, 4, 2, 1. Note that mbcd complement is nines complement of decimal digit.

Stored internally as 17 mbcd characters + 3 blank characters, or as 80 bits.

Transfers are serial between decimal digits, parallel within decimal digits.

D. INSTRUCTION WORD FORMAT

Three addresses plus operation code.

Address A, operation code, address B, address C.

Addresses A and B each composed of transfer sign, channel no., no. time—2 bits, 6 bits, 4 bits respectively. Total bits 12 each. (or 1 "quad", 1 "quad" + 1 decimal, 1 decimal)

Address C similar, but no transfer sign, 10 bits total.

Operations code, 4 bits, mbcd.

Each instruction thus takes 38 bits.

E. OPERATING TIMES

One cycle = 1/250 sec. or 4 millisec.

Access time—fast storage 4 millisec.

Access time—slow storage[1]

Addition—4 millisec., 1 cycle

Multiplication—12 millisec., 3 cycles

Built-in functions (data not available)

Read/write one data word on magnetic tape 0.1 sec.

Printing—3.5 sec. for 16 decimal digit word

F. SOURCE OF KEY COMPONENTS

Magnetic tape and magnetic drum units and electronic circuitry designed and built by Computation Laboratory.

G. CHRONOLOGY

1946 Initial research and conceptual work begun

1948 (summer) Started detailed design and construction

1951 (March)[2] Moved to Dahlgren Naval Proving Ground (now the Naval Weapons Laboratory) after assembly and test at Harvard

1952 (January)[3] Started production operations

1. From slow to fast storage, transfer groups of 20 data words. From fast to slow storage, transfer groups of 10 data words.
2. These are dates given by Dahlgren; dates given in Harvard Manual of Operation (*Annals*, vol, 25) are considerably earlier.
3. These are dates given by Dahlgren; dates given in the Manual of Operation (*Annals*, volume 25) are considerably earlier.

1955 Retired from use, some time after delivery of NORC computer to Dahlgren

MARK IV FACT SHEET 1[4]

A. FUNCTIONAL EQUIPMENT

Units of Mark IV are divided into two groups, called beta and gamma. Transfers of numbers are from beta to gamma or gamma to beta (under control of alpha-code). Numbers can also be stored on magnetic drum (slow storage) which also provides storage for all coding.

On the beta side

230 fast storage registers (magnetic core shift registers) of which 20 can be used to store and recall numbers from the drum

1 shift and transfer register

1 shift control register

2 index registers (one for fast storage, one for tape units)

1 magnetic tape read register

1 magnetic tape record register

1 line number register (contains number of lines of coding being executed)

1 condition register (for IF calls)

1 call register

4. Data are taken principally from Kenneth Iverson's report in Progress Report No. 23, May-August 1952. The magnetic drum is described by R. Hofheimer and R. Wilkins in Progress Report No. 19, August-November 1951. There is no officially published manual for Mark IV. There is, however, a Manual of Operation for the Harvard Magnetic Drum Calculator (Mark IV) (reproduced from a typed copy by some form of photolithographic reproduction), written by Norman B. Solomon and dated July 1957.

1 slow storage control register

1 sign register

1 decimal point register

(the last two are manually set)

On the gamma side

2 accumulators share logic circuitry; can be ganged together as a double precision accumulator

1 multiply unit

1 divide unit

1 transfer register

On the drum

4,000 slow-access storage registers (for numbers)

10,000 lines of coding

Software (usually kept on the drum)

double precision arithmetic

complex arithmetic

real functions: $x^{1/2}$, $\cos x$, $\tan^{-1}$s, $\log_{10} x$, 10^x

routines to test the operation of components of the computer

Input/Output

12 magnetic tape units

A set of manual controls

1 set of indications to display the contents of a register

1 set of indications to display a line of coding

Off-line

1 unit to record data on magnetic tap (manually)

1 unit to record instructions on magnetic tape (manually)

B. PRINCIPAL COMPONENTS

electronic tubes—mostly glass envelope, miniature

selenium diodes, replaced by germanium (Raytheon) diodes for nearly all logic. Arranged on cards made at Harvard.

over 30,000 magnetic cores for fast storage.

1 Magnetic Drum designed at Harvard, fabricated by Pratt & Whitney

Diameter 22 inches

Usable height 20 inches

Pulse density 29 pulses/inch

Speed of rotation 1800 r.p.m.

Number of tracks 274

12 vacuum drive magnetic tape units with photoelectrically controlled servomotor drives for tape reels—designed at Harvard

motor generator sets for D.C. power

cartridge fuses by the score

Indicator dials were of type used to read out stock market quotations. We adapted them for our use.

C. DATA WORD FORMAT

16 decimal digits & sign. Each decimal digit represented by four bits.

Decimal point adjustable by switch on control panel

Decimal point 15 available always (irrespective of switch setting) by special codes for multiply and divide

Internal transfers are series parallel, 1 decimal digit at a time

D. INSTRUCTION FORMAT

20 bits divided into alpha, beta, and gamma codes

alpha (4 bits) determines direction and sign of number transfer

beta (10 bits) and gamma (6 bits) codes determine beta and gamma addresses

However, special alpha codes (04 & 05) followed by 4-decimal digit number cause that number to be placed in gamma transfer register. This permits constants to be introduced from the coding.

E. OPERATING TIME

Each line of coding is executed in 1.2 ms.

Clear accumulator and read a number in takes 1.2 ms

Add a number to sum stored in accumulator 1.2 ms

Read out sum 1.2 ms

Multiplication 12 ms

Division 26.4 ms

Note: while multiplication and/or division are in progress, other interposed commands are executed

Transfer 10 numbers between slow & fast storage 44.4 ms (avg)

Read one number from magnetic tape 26.4 ms

Record one number on magnetic tape 26.4 ms

Transfer of control (call, GO TO) 2.4–31.2 ms

Aiken's Doctoral Students and Their Dissertations

The first two doctorates (Mitchell and Hayes) were officially awarded in Engineering Sciences and Applied Physics; all the rest were in Applied Mathematics.

1948
Herbert Francis Mitchell Jr., "The Application of Large-Scale Digital Calculators to the Solution of Simultaneous Linear Systems"

1950
Miles Van Valzah Hayes, "Numerical Solutions of Differential Equations, Using Automatic Computers"

1952
Gerrit Anne Blaauw, "The Application of Selenium Rectifiers as Switching Devices in the Mark IV Calculator"

1953
Charles Allerton Coolidge Jr., "Design of an Automatic Digital Calculator"

Robert Charles Minnick, "The Use of Magnetic Cores as Switching Devices"

1954
Kenneth Eugene Iverson, "Machine Solutions of Linear Differential Equations: Applications to a Dynamic Economic Model"

Anthony Gervin Oettinger, "A Study for the Design of an Automatic Dictionary"

Warren Lloyd Semon, "The Application of Matrix Methods in the Theory of Switching"

Theodore Singer, "A Class of Time-Sequential Circuits"

1955
Peter Calingaert, "Multiple-Output Relay Switching Circuits"

1956
Robert Lovett Ashenhurst, "The Structure of Multiple-Coincidence Selection Systems"

Frederick Phillips Brooks Jr., "The Analytic Design of Automatic Data Processing Systems"

LeRoy Brown Martin Jr., "Approximation by Ratios of Integers Chosen from a Limited Range-Applications to the Gear Train Problem"

1957
Roderick Gould, "The Application of Graph Theory to the Synthesis of Contact Networks"

Albert Lafayette Hopkins Jr., "An Investigation of Non-Ohmic Resistive Switching Networks"

1958
Gerard A. Salton, "An Automatic Data Processing System for Public Utility Revenue Accounting"

Index